C N C

Technology

and

Programming

Steve Krar

Arthur Gill

Gregg Division
McGraw-Hill Publishing Company
New York   Atlanta   Dallas   St. Louis   San Francisco
Auckland   Bogotá   Caracas   Hamburg   Lisbon
London   Madrid   Mexico   Milan   Montreal   New Delhi
Paris   San Juan   São Paulo   Singapore
Sydney   Tokyo   Toronto

Sponsoring Editor: Steve Zollo
Design and Art Supervisors: Janice Noto and Meri Shardin
Production Supervisor: Mirabel Flores

Interior and Cover Designer: Robin Hessel Hoffman
Cover Illustrator: Noi Viva Design
Technical Art: York Production Services

Library of Congress Cataloging-in-Publication Data

Krar, Stephen F.
    CNC : technology and programming / Steve Krar, Arthur Gill.
       p.   cm.
    Includes index.
    ISBN 0-07-023333-0 : $29.95
    1. Machine-tools—Numerical control. I. Gill, Arthur, date.
  II. Title.
  TJ1189.K74 1990
  621.9′023—dc20   89-12571
                                         CIP

Some of the artwork for this book was processed electronically.

CNC: Technology and Programming

1 2 3 4 5 6 7 8 9 0 DOCDOC 8 9 6 5 4 3 2 1 0 9

ISBN 0-07-023333-0

Contents

Preface

The term *numerical control* is a widely accepted and commonly used term in the machine tool industry. Numerical control (NC) enables an operator to communicate with machine tools through a series of numbers and symbols.

NC and computer numerical control (CNC) have brought tremendous changes to the metalworking industry. New machine tools in combination with NC or CNC have enabled industry to consistently produce parts to accuracies undreamed of only a few years ago. The same part can be reproduced to the same degree of accuracy any number of times if the NC tape has been properly prepared or if the computer has been properly programmed. The operating commands which control the machine tool are executed automatically with amazing speed, accuracy, efficiency, and repeatability.

The ever-increasing use of NC and CNC in industry has created a need for personnel who are knowledgeable about and capable of preparing the programs which guide the machine tools to produce parts to the required shape and accuracy. With this in mind, the authors have prepared this textbook to take the mystery out of NC—to put it into a logical sequence and express it in simple language that everyone can understand. The preparation of a program is explained in a logical step-by-step procedure, with practical examples to guide the student. Review questions at the end of each unit are designed to stimulate reasoning and to increase knowledge of this exciting and ever-expanding field.

The book begins with the history of NC, to give the reader some background on how and why it was developed. Among the subjects covered are:

1. The types of input media (what commands the machine tool to perform operations)

2. Simple and contour programming

3. Examples of three machine tools (machining centers, turning centers, electrodischarge machines) that use NC

4. Brief coverage of CNC and the effect it is having on the metalworking industry

5. The future and the various developments that will play an important role in the machine tool trade

It is advisable for a person who wishes to follow a career in NC to have some background in machining fundamentals, in order to know how to program the sequence of operations required to produce a finished part. An introductory machine shop course covering the basic machining fundamentals and sequences on various machine tools would also certainly be an asset to a person learning to become an NC programmer. Although this knowledge is desirable, however, it is not a prerequisite to learning about NC.

NC plays a significant role in the overall manufacturing facility along with computer-aided design (CAD), computer-aided manufacturing (CAM), computer-integrated manufacturing (CIM), and flexible manufacturing systems (FMS). The advances in microelectronics and computers are truly revolutionizing manufacturing.

Steve Krar
Arthur Gill

Acknowledgments

The authors wish to express their sincere appreciation to Alice Krar for the many hours spent typing, proofreading, and checking the manuscript. Her assistance was of prime importance in making the book as clear as possible for the learner and teacher alike.

We owe a special debt of gratitude to the many students, teachers, and industrial personnel who reviewed sections of the book and offered constructive criticism and suggestions for improving this text. A special note of thanks goes to the students of the Drafting Department at Niagara College, especially Larry Strong, for their assistance with the many line drawings which appear in the book.

We also greatly appreciate the assistance of the following firms, which were kind enough to review sections of the manuscript and supply technical information and illustrations for this book.

Agie USA, Ltd.
Allen-Bradley Company
American SIP Corporation
Bausch & Lomb
Butterfield Division, Union Twist Drill
Cincinnati Milacron, Inc.
Cleveland Twist Drill Company
Coleman Engineering Company, Inc.
Electronic Industries Association
Elox Division, Colt Industries
Facit, Inc.
General Electric Company
Greenfield Tap and Die Company
Hertel Carbide Ltd.
Hewlett Packard
Hitachi-Seiki U.S.A. Inc.
Icon Corporation

Kennametal Inc.
Kostel Enterprises Ltd.
LeBlond Makino Machine Tool Company
Mazak Corporation
Modern Machine Shop
Moore Special Tool Company
Northwestern Tools, Inc.
Numeridex Inc.
Pederson Company
Philips Electronic Instruments
PowerHold Inc.
Rockwell International
Standard Tool Company
Starrett, L.S. Company
Superior Electric Company
Taft-Peirce Mfg. Company
Weldon Tool Company
J.H. Williams & Company

CHAPTER

ONE

History of Numerical Control

Numerical control (NC) is the operation of a machine tool by a series of coded instructions consisting of numbers, letters of the alphabet, and symbols which the machine control unit (MCU) can understand. These instructions are converted into electrical pulses of current which the machine's motors and controls follow to carry out machining operations on a workpiece. The numbers, letters, and symbols are coded instructions which refer to specific distances, positions, functions, or motions which the machine tool can understand as it machines the workpiece.

After completing
this chapter,
you should be
able to:

1. Understand the importance of NC in the metalworking industry

2. Know and understand the Cartesian coordinate measuring system

3. Identify three of the more common machine tools using NC

4. Know the difference between the absolute and incremental systems of measurement

HISTORY

A form of NC was used in the early days of the industrial revolution, as early as 1725, when knitting machines in England used punched cards to form various patterns in cloth. Even earlier than this, rotating drums with prepositioned pins were used to control the chimes in European cathedrals and some American churches. In 1863, the first player piano was patented; it used punched paper rolls, through which air passed, to automatically control the order in which the keys were played (Fig. 1-1).

The principle of mass production (interchangeable manufacture), developed by Eli Whitney, transferred many operations and functions originally performed by skilled artisans to the machine tool. As better and more precise machine tools were developed, the system of interchangeable manufacture was quickly adopted by industry in order to produce large quantities of identical parts. In the second half of the nineteenth century, a wide range of machine tools were developed for the basic metal-cutting operations, such as turning, drilling, milling, and grinding. As better hydraulic, pneumatic, and electronic controls were developed, better control over the movement of machine slides became possible.

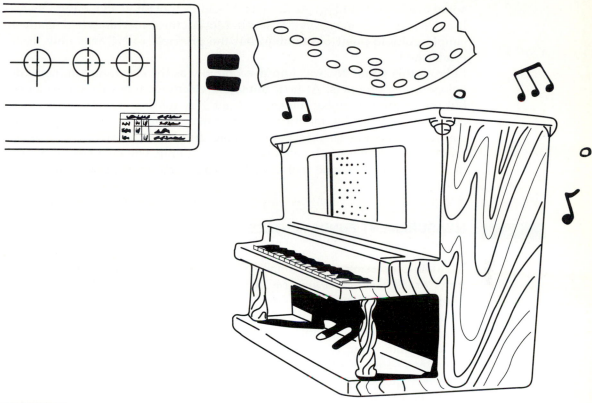

Fig. 1-1   The player piano used air passing through holes in a drum to operate the keys in a programmed sequence. (*Coleman Engineering Company*)

In 1947, the U.S. Air Force found that the complex designs and shapes of aircraft parts—such as helicopter rotor blades and missile components—were causing problems for manufacturers, who could not keep up to projected production schedules. At this time, John Parsons, of the Parsons Corporation, of Traverse City, Michigan, began experimenting with the idea of making a machine tool generate a "thru-axis curve" by using numerical data to control the machine tool motions. In 1949, the U.S. Air Matériel Command awarded Parsons a contract to develop NC and in turn speed up production methods. Parsons subcontracted this study to the Servomechanism Laboratory of the Massachusetts Institute of Technology (MIT), which in 1952 successfully demonstrated a vertical spindle Cincinnati Hydrotel, which made parts through simultaneous three-axis cutting tool movements. In a very short period of time, almost all machine tool manufacturers were producing some machines with NC. At the 1960 Machine Tool Show in Chicago, over a hundred NC machines were displayed. Most of these machines had relatively simple point-to-point positioning, but the principle of NC was now firmly established.

From this point, NC was improved rapidly as the electronics industry developed new products. At first, miniature electronic tubes were developed, but the controls were big, bulky, and not very reliable. Then solid-state circuitry and, eventually, modular, or integrated, circuits were developed. The control unit became smaller, more reliable, and less expensive. The development of even better machine tools and control units will help spread the use of NC from the machine tool industry to all facets of manufacturing.

MEASUREMENT FUNDAMENTALS

NC data processing (with numbers, letters, and symbols) is done in a computer or *machine control unit* (MCU) by adding, subtracting, multiplying, dividing, and comparing. The computer can be programmed to recognize an A command before a B command, an item 1 before an item 2, or any other elements in their sequential order. Subtracting is done by adding negative values, multiplying is done through a series of additions, and dividing is done by a series of subtractions. The computer is capable of handling numbers very quickly; the addition of two simple numbers may take only a billionth of a second (a nanosecond).

Binary Numbers

Primitive people used their 10 fingers and 10 toes to count numbers, and from this evolved our present decimal, or Arabic, system, where "base ten," or the power of 10, is used to signify a numerical value. Computers and MCUs, in contrast, use the binary, or base 2, system to recognize numerical values. A knowledge of the binary system is not essential for the operator since both the computer and the MCU can recognize the standard decimal system and convert it into binary data.

To provide an understanding of the binary system, let us compare it with the decimal system.

1. *Decimal system*
   In the decimal system the value of each digit depends on where it is placed in relation to the other digits in a number. The number one (1) by itself is worth 1, but if it is placed to the left of one or two zeros (0),

it is worth 10 and 100, respectively. Before numbers can be added or subtracted, they must first be arranged in their proper place columns. In the decimal system, each position to the left of a decimal point means an increase in the power of 10.

2. *Binary system*
The binary system uses only two digits, zero (0) and one (1), and is based on the power of two (2) instead of ten (10), as in the decimal system. Each position to the left means an increase in the power of two (2). For example:

$$2^1 = 2$$
$$2^2 = 4$$
$$2^3 = 8 \ (2 \times 2 \times 2)$$
$$2^4 = 16 \ (2 \times 2 \times 2 \times 2)$$
$$2^5 = 32$$

Therefore, any numerical value can be represented using only the two digits, one (1) and zero (0). Since there are only two digits, the binary system is often called the "ON-or-OFF" system. For example:

$$1 = ON$$
$$0 = OFF$$

On numerical punched tape, a hole represents a one (1) and no hole represents a zero (0).

Table 1-1 illustrates the differences between the decimal and binary systems. Examples are shown to clarify both systems.

The rules of addition in the binary system are somewhat different from those of the decimal system because in the binary system only two digits are involved. They are explained in Table 1-2. See Table 1-3 for a comparison of addition in the decimal and binary systems.

Binary numbers are essential for the computer and the MCU, which process information at a very high speed. Binary notations are used because electrical circuits are stable in either of two conditions: ON or OFF, POSITIVE or NEGATIVE, CHARGED or DISCHARGED, and so forth. Since a punched tape has either a hole or no hole at a specific position, the tape reader on a machine tool decodes this information and converts it into electrical pulses.

Computers and MCUs accept information in the decimal system commonly used in industry and convert this information into binary form. The programmer does not have to make any conversions from one system to the other.

Table 1-1   **COMPARISON OF THE DECIMAL AND BINARY SYSTEMS**

Decimal system (Power of 10)		Binary System (Power of 2)			
Place 10	Place 1	Place 8	Place 4	Place 2	Place 1
0					0
1					1
2				1	0
3				1	1
4			1	0	0
5			1	0	1
6			1	1	0
7			1	1	1
8		1	0	0	0
9		1	0	0	1
1	0	1	0	1	0

To convert the number 55 from decimal to binary

1. Subtract the largest possible power of two from the number
2. Mark a 1 (one) in each column used
3. Keep subtracting the largest possible power of two from the remainder
4. Mark a 0 (zero) in each column not used
5. Keep subtracting (power of two) until the remainder is 0 (zero)

Example:

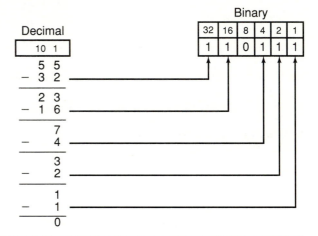

Table 1-2   **BINARY ADDITION RULES**

0 +0 ―― 0	0 +1 ―― 1	1 +1 ―― 1 0
Same as the decimal system	Same as the decimal system	1 + 1 = 0 with 1 carried to the next column left

Table 1-3	COMPARISON OF ADDITION IN THE DECIMAL AND BINARY SYSTEMS

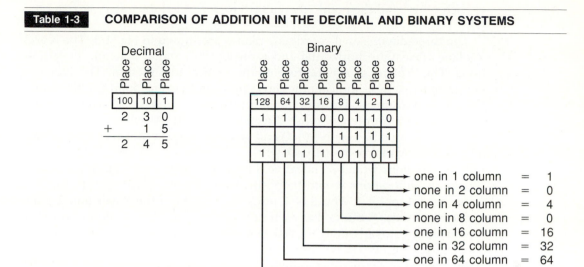

CARTESIAN COORDINATE SYSTEM

Almost everything which can be produced on a conventional machine tool can be produced on a numerical control machine tool, with its many advantages. The machine tool movements used in producing a product are of two basic types: *point-to-point* (straight-line movements) and *continuous path* (contouring movements).

The Cartesian, or rectangular, coordinate system was devised by the French mathematician and philosopher René Descartes. With this system, any specific point can be described in mathematical terms from any other point along three perpendicular axes. This concept fits machine tools perfectly since their construction is generally based on three axes of motion (X, Y, Z) plus an axis of rotation. On a plain vertical milling machine, the X axis is the horizontal movement (right or left) of the table, the Y axis is the table cross movement (toward or away from the column), and the Z axis is the vertical movement of the knee or the spindle. NC systems rely heavily on the use of rectangular coordinates because the programmer can locate every point on a job precisely.

When points are located on a workpiece, two straight intersecting lines, one vertical and one horizontal, are used. These lines must be at right angles to each other, and the point where they cross is called the *origin*, or *zero point* (Fig. 1-2).

The three-dimensional coordinate planes are shown in Fig. 1-3. The X and Y planes (axes) are horizontal and represent horizontal machine table motions. The Z plane or axis represents the vertical tool motion. The Plus (+) and minus (−) signs indicate the direction from the zero point (origin) along the axis of movement. The four quadrants formed when the XY axes cross are numbered in a counterclockwise direction (Fig. 1-4). All positions located in quadrant 1 would be positive (+X) and positive (+Y). In the second quadrant, all positions would be negative X (−X) and positive (+Y). In the third quadrant, all locations would be negative X (−X) and negative (−Y). In the fourth quadrant, all locations would be positive X (+X) and negative Y (−Y).

In Fig. 1-4, point A would be 2 units to the right of the Y axis and 2 units above the X axis. Assume that each unit equals 1 in. The location of point A would be X + 2.000 and Y + 2.000. For point B, the location would be X + 1.000 and Y − 2.000. In NC programming it is not necessary to indicate plus (+) values since these are assumed. However, the minus (−) values must be indicated. For example, the locations of both A and B would be indicated as follows:

```
A  X2.000  Y2.000
B  X1.000  Y - 2.000
```

Intersecting lines form right angles and establish the zero point. (*Allen-Bradley*)

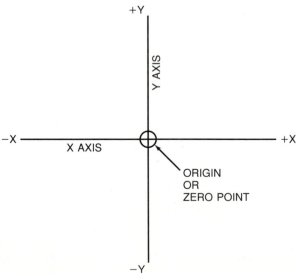

Fig. 1-3 The three-dimensional coordinate planes (axes) used in NC. (*The Superior Electric Company*)

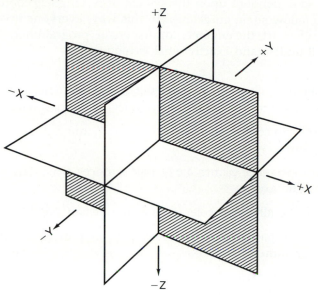

Fig. 1-4 The quadrants formed when the X and Y axes cross are used to accurately locate points from the XY zero, or origin, point. (*Allen-Bradley*)

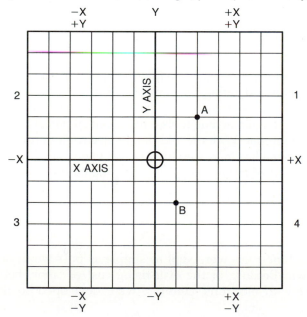

Guidelines

Since NC is so dependent upon the system of rectangular coordinates, it is important to follow some guidelines. In this way, everyone involved in the manufacture of a part, the engineer, draftsperson, programmer, and machine operator, will understand exactly what is required.

1. Use reference points on the part itself, if possible. This makes it easier for quality control to check the accuracy of the part later.

2. Use Cartesian coordinates—specifying X, Y, and Z planes—to define all part surfaces.

3. Establish reference planes along part surfaces which are parallel to the machine axes.

4. Establish the allowable tolerances at the design stage.

5. Describe the part so that the cutter path may be easily determined and programmed.

6. Dimension the part so that it is easy to determine the shape of the part without calculations or guessing.

MACHINES USING NC

Early machine tools were designed so that the operator was standing in front of the machine while operating the controls. This design is no longer necessary, since in NC the operator no longer controls the machine tool movements. On conventional machine tools, only about 20 percent of the time was spent removing material. With the addition of electronic controls, actual time spent removing metal has increased to 80 percent and even higher. It has also reduced the amount of time required to bring the cutting tool into each machining position.

Machine Types

In the past, machine tools were kept as simple as possible in order to keep their costs down. Because of the ever-rising cost of labor, better machine tools, complete with electronic controls, were developed so that industry could produce more and better goods at prices which were competitive with those of offshore industries.

NC is being used on all types of machine tools, from the simplest to the most complex. The most common machine tools are the single-spindle drilling machines, lathe, milling machine, turning center, and machining center.

1. *Single-spindle drilling machine*
   One of the simplest numerically controlled machine tools is the *single-spindle drilling machine* (Fig. 1-5). Most drilling machines are programmed on three axes:

   a. The X axis controls the table movement to the right and left.

   b. The Y axis controls the table movement toward or away from the column.

   c. The Z axis controls the up or down movement of the spindle to drill holes to depth.

Fig. 1-5   A numerically controlled single-spindle drilling machine showing the X, Y, and Z axes. (*Cincinnati Milacron, Inc.*)

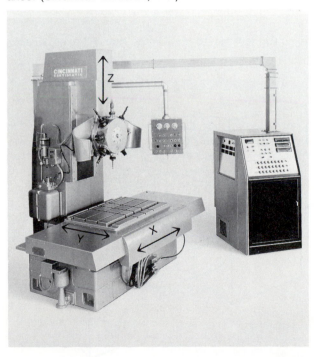

2. *Lathe*

The *engine lathe*, one of the most productive machine tools, has always been a very efficient means of producing round parts (Fig. 1-6). Most lathes are programmed on two axes:

a. The X axis controls the cross motion (in or out) of the cutting tool.

b. The Z axis controls the carriage travel toward or away from the headstock.

3. *Milling Machine*

The *milling machine* has always been one of the most versatile machine tools used in industry (Fig. 1-7). Operations such as milling, contouring, gear cutting, drilling, boring, and reaming are only a few of the many operations which can be performed on a milling machine. The milling machine can be programmed on three axes:

a. The X axis controls the table movement left or right.

b. The Y axis controls the table movement toward or away from the column.

c. The Z axis controls the vertical (up or down) movement of the knee or spindle.

Fig. 1-6 The engine lathe cutting tool moves only on the X and Z axes. (*Electronic Industries Association*)

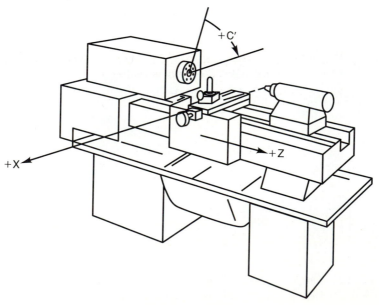

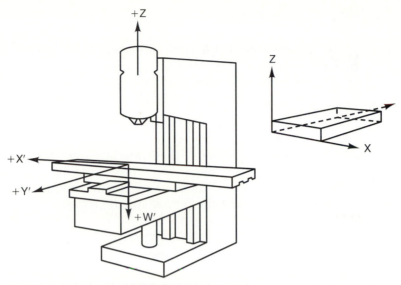

Fig. 1-7 The vertical knee and column milling machine operates on the X, Y, and Z axes. (*Electronic Industries Association*)

4. *Turning Center*

Turning centers (Fig. 1-8) were developed in the mid-1960s after studies showed that about 40 percent of all metal cutting operations were performed on lathes. These numerically controlled machines are capable of greater accuracy and higher production rates than were pos-

Fig. 1-8 Turning centers are capable of producing round work to a high degree of accuracy and with high production rates. (*Cincinnati Milacron, Inc.*)

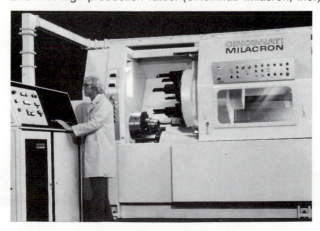

sible on the engine lathe. The basic turning center operates on only two axes:

   a. The X axis controls the cross motion of the turret head.

   b. The Z axis controls the lengthwise travel (toward or away from the headstock) of the turret head.

5. *Machining Center*

Machining centers (Fig. 1-9) were developed in the 1960s so that a part did not have to be moved from machine to machine in order to perform various operations. These machines greatly increased production rates because more operations could be performed on a workpiece in one setup. There are two main types of machining centers, the *horizontal* and the *vertical* spindle types.

Fig. 1-9   Machining centers are capable of performing a variety of machining operations on a workpiece. (*Cincinnati Milacron, Inc.*)

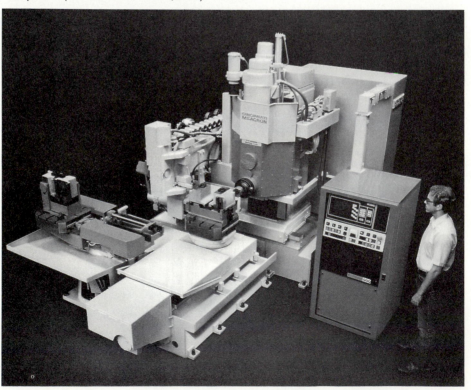

Fig. 1-10   Horizontal spindle machining centers have three axes of motion. (*Bridgeport Machines Division of Textron, Inc.*)

    a. The horizontal spindle machining center (Fig. 1-10) operates on three axes:
      i. The X axis controls the table movement left or right.
      ii. The Y axis controls the vertical movement (up or down) of the spindle.
      iii. The Z axis controls the horizontal movement (in or out) of the spindle.
    b. The vertical spindle machining center (Fig. 1-11) operates on three axes:
      i. The X axis controls the table movement left or right.
      ii. The Y axis controls the table movement toward or away from the column.
      iii. The Z axis controls the vertical movement (up or down) of the spindle.

Fig. 1-11 Vertical spindle machining centers have three axes of motion. (*Cincinnati Milacron, Inc.*)

PROGRAMMING SYSTEMS

Two types of programming modes, the incremental system and the absolute system, are used for NC. Both systems have applications in NC programming, and no system is either right or wrong all the time. Most controls on machine tools built today are capable of handling either incremental or absolute programming.

Incremental System

In the incremental system, dimensions or positions are given from a previous known point. As an example of incremental instructions, consider a person who delivers newspapers to certain houses on a street. He or she could be given instructions to deliver a newspaper to the first house, 60 ft (feet) from

REFERENCE POINT SYSTEMS

```
┌─────────────────────────────┐
│        POSITIONING          │
│  REFERENCE POINT SYSTEMS    │
└─────────────────────────────┘
       ↙               ↘
┌──────────────┐   ┌──────────────┐
│ INCREMENTAL  │   │   ABSOLUTE    │
└──────────────┘   └──────────────┘
```

the corner (Fig. 1-12). The second house which should have a paper could then be described as 120 ft from the first house, and the third house as 60 ft from the second. Note that all distances are expressed from the known previous point. Incremental dimensioning on a job print is shown in Fig. 1-13. As you will note, the dimensions for each hole are given from the previous hole. One disadvantage of incremental positioning or programming is that if there is an error made in any location, this error is carried over to all the locations made after this point.

Incremental program locations are always given as the distance and direction from the immediately preceding point (Fig. 1-13). Command codes which tell the machine to move the table, spindle, and knee are explained here using a vertical milling machine as an example:

- A "plus X" (+X) command will cause the cutting tool to be located to the right of the last point.

Fig. 1-12 The incremental system being used to locate the houses that require newspapers. (*Coleman Engineering Company*)

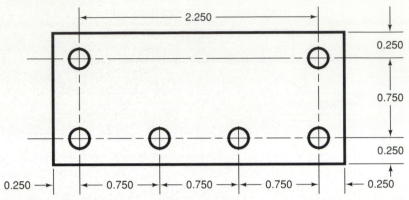

2.250

0.250

0.750

0.250

0.250   0.750   0.750   0.750   0.250

Fig. 1-13 A workpiece dimensioned in the incremental system mode. (*Icon Corporation*)

- A "minus X" (−X) command will cause the cutting tool to be located to the left of the last point.
- A "plus Y" (+Y) command will cause the cutting tool to be located toward the column.
- A "minus Y" (−Y) will cause the cutting tool to be located away from the column.
- A "plus Z" (+Z) command will cause the cutting tool or spindle to move up or away from the workpiece.
- A "minus Z" (−Z) moves the cutting tool down or into the workpiece.

In incremental programming, the G91 command will indicate to the computer and MCU that programming is to be in the incremental mode.

Absolute System

In the absolute system, all dimensions or positions are given from a zero or reference point. For instance, consider the person delivering newspapers but now using absolute instructions and using the street corner as the zero or reference point. The first house is 60 ft from the corner (Fig. 1-14), the second house is 180 ft from the corner, the third house is 240 ft from the corner, and so on. As you will note, all distances have been given from the corner, which is the zero or reference point. In Fig. 1-15, the same workpiece is used as in Fig. 1-13, but all dimensions are given from the zero or reference point. Therefore, in the absolute system of dimensioning or programming, an error in any dimension is still an error, but the error is not carried on to any other location.

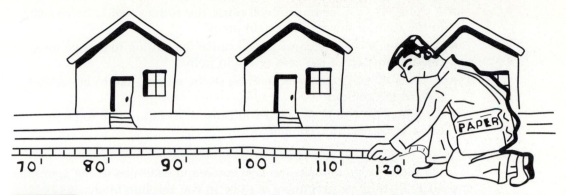

Fig. 1-14 The absolute system located all houses requiring a newspaper by the distance each is from the street corner. (*Coleman Engineering Company*)

Absolute program locations are always given from a single fixed zero or origin point (Fig. 1-15). The zero or origin point may be a position on the machine table, such as the corner of the worktable or at any specific point on the workpiece. In absolute dimensioning and programming, each point or location on the workpiece is given as a certain distance from the zero or reference point.

Fig. 1-15 A workpiece dimensioned in the absolute system mode. *Note:* All dimensions are given from a known point of reference. (*Icon Corporation*)

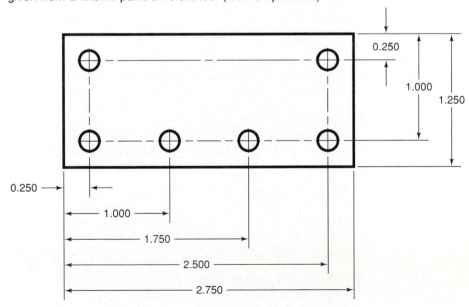

- A "plus X" (+X) command will cause the cutting tool to be located to the right of the zero or origin point.
- A "minus X" (−X) command will cause the cutting tool to be located to the left of the zero or origin point.
- A "plus Y" (+Y) command will cause the cutting tool to be located toward the column.
- A "minus Y" (−Y) command will cause the cutting tool to be located away from the column.

In absolute programming, the G90 command indicates to the computer and MCU that the programming is to be in the absolute mode.

ADVANTAGES OF NC

NC has grown at an ever-increasing rate, and its use will continue to grow because of the many advantages that it has to offer industry. Some of the most important advantages of NC are illustrated in Fig. 1-16.

1. *Greater Operator Safety*
   NC systems are generally operated from a console which is usually away from the machining area; therefore the operator is not exposed to moving machine parts or to the cutting tool.

2. *Greater Operator Efficiency*
   Since an NC machine does not require much attention, the operator can perform other jobs while the machine is running.

3. *Reduction of Scrap*
   Because of the high degree of accuracy of NC systems and the elimination of most of the human errors, scrap has been drastically reduced.

4. *Reduced Lead Time for Production*
   The tape preparation and setup for numerically controlled machines is usually short. Many jigs and fixtures formerly required are not necessary. The tape can be filed in a small space and used again for future production runs.

5. *Reduction of Human Error*
   The NC tape eliminates the need for an operator to take trial cuts, make trial measurements, or make table positioning movements.

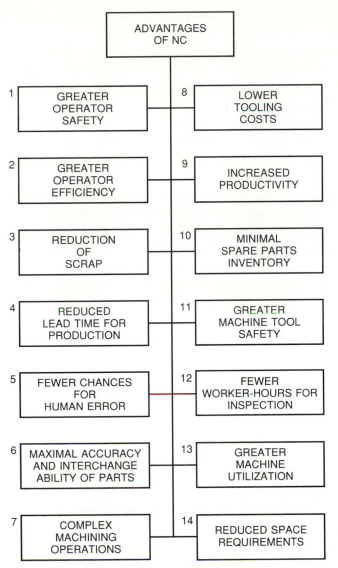

Fig. 1-16   Advantages of NC in the industrial world.

There is no longer a need for the operator to change cutting tools, select the sequence of operations, and perform other routine functions.

6. *High Degree of Accuracy*
NC ensures that all parts produced will be accurate and of uniform

quality. The improvement of accuracy of the parts produced by NC assures the interchangeability of parts.

7. *Complex Machining Operations*
Complex operations can be done more quickly and accurately with NC and electronic measuring equipment.

8. *Lower Tooling Costs*
NC generally requires simple holding fixtures; therefore, the cost of fixture design and manufacture may be reduced by as much as 70 percent.

9. *Increased Productivity*
Because the NC controls all the machine functions, parts are produced faster, with less setup and lead time.

10. *Reduced Parts Inventory*
A large inventory of spare parts is no longer necessary, since additional parts can be made to the same accuracy when the same tape is used.

11. *Greater Machine Tool Safety*
The damage to machine tools as a result of operator error is virtually eliminated because there is less need for operator intervention.

12. *Less Inspection Required*
Because NC produces parts of uniform quality, less inspection time is required. Once the first part has passed inspection, very little additional inspection is required.

13. *Greater Machine Use*
Because there is less time required for setup and operator adjustments, production rates could increase as much as 80 percent.

14. *Reduced Space Requirements*
NC requires fewer jigs and fixtures and therefore less storage space.

REVIEW QUESTIONS

1. Define *numerical control.*

History

2. Name three early forms of NC.

3. Who first used numerical data to control machine tool motions?

4. How did the development in the electronics industry spur the use of NC in the machine tool industry?

Measurement Fundamentals

5. Compare the decimal system with the binary system.

6. Why are only two numbers used in the binary system, and how do they affect electrical circuits?

Cartesian Coordinate System

7. Name the two basic classifications of positioning.

8. Why does the Cartesian coordinate system fit machine tools perfectly?

9. Using a vertical milling machine, define the X, Y, and Z axes.

10. Make a sketch of the four quadrants and explain what each represents.

11. Why is it important in NC work that some guidelines be used with the system of rectangular coordinates?

Machines Using NC

12. How has the addition of electronic controls affected the amount of time spent on removing metal?

13. Name four of the most common machine tools which use NC.

14. What machine tool is programmed on only two axes?

Programming Systems

15. Compare the incremental and absolute systems with regard to dimensioning.

16. Define the zero, or origin, point.

17. What do the $+X$, $-X$, $+Y$, $-Y$ commands represent in:
   (a) Incremental mode
   (b) Absolute mode

Advantages of NC

18. List four of the most important advantages of NC for each of the following:
   (a) Production rates
   (b) Part accuracy
   (c) Cost reduction

CHAPTER

TWO

The Computer

Throughout history, the development of various tools and machines has had a profound effect on the lifestyles and standards of living throughout the world. Humanity has progressed through the Stone Age, Bronze Age, Iron Age, and Machine Age, and each has helped to improve humanity's lot and make it more productive. We are now in the Computer Age, and nothing in our past history can compare to the effect that the computer is having on the way we live, work, and play. It will make us more productive, provide us with more leisure time, remove the drudgery of home chores, and in general radically change the way humanity lives.

After completing
this chapter,
you should be
able to:

1. Describe the development of the computer from primitive times to the present

2. Summarize the effect computers are having on manufacturing and everyday life

HISTORY OF THE COMPUTER

Ever since recorded history, people have used some device to count and perform calculations. Primitive people used things such as their fingers and toes or stones to count (Fig. 2-1). The world's first computer was developed in the Orient around 4000 B.C. It was the abacus (Fig. 2-2), and it involved the moving of beads on several wires to count numbers. The abacus is a very accurate instrument when properly used and may still be found in use in some Oriental businesses.

Fig. 2-1 The methods used by primitive people to make calculations.

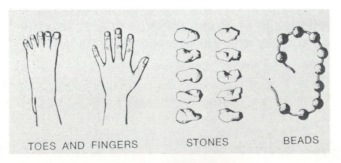

TOES AND FINGERS   STONES   BEADS

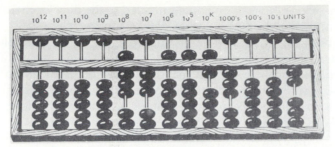

Fig. 2-2 The abacus, developed around 4000 B.C., was the world's first real computer.

In 1642, the first mechanical calculator was constructed by a Frenchman named Blaise Pascal. It consisted of eight wheels or dials, each with the numbers 0 to 9. Each wheel represented units, 1s, 10s, 100s, 1000s, and so forth. It could, however, only add or subtract. Multiplication or division was done by repeated additions or subtractions.

In 1671, a German mathematician added the capability of multiplication and division. However, this advanced machine could only do arithmetical problems.

In the nineteenth century, Charles Babbage, an English mathematician, produced a machine, called the *difference engine*, that could rapidly and accurately calculate long lists of various functions, including logarithms.

In 1804, a French mechanician, J. M. Jacquard, introduced a punch-card system to direct the operations of a weaving loom. In the United States, Herman Hollerith introduced the use of punched cards to record personal information, such as age, sex, race, and marital status, for the 1890 U.S. Census (Fig. 2-3). The information was encoded on cards and read and tabulated by

Fig. 2-3 Punched cards, the first method of data processing, were used as early as 1890.

electric sensors. This use of punched cards led to the development of the early office machines for the tabulation of data.

In the 1930s, a German named Konrad Zuse built a simple computer which, among other things, was used to calculate wing designs for the German aircraft industry. In 1939, a mathematician named George Stibitz produced a similar device for the Bell Telephone Laboratories in the United States. This machine was capable of doing calculations over telephone wires; thus was born the first remote data processing machine.

During World War II, the British built a computer called the Colossus I, which helped break the German military codes. The earliest digital computers used electromechanical ON-OFF switches or relays. The first large computer, the Mark I, assembled at Harvard University by IBM, could multiply two 23-digit numbers in about 5 seconds—a very slow feat compared to today's machines.

In 1946, the world's first electronic digital computer, the ENIAC (Electronic Numerical Integrating Automatic Computer), was produced. It contained more than 19,000 vacuum tubes, weighed almost 30 tons, and occupied more than 15,000 ft^2 (square feet) of floor space. It was a much faster computer than any that preceded it, able to add two numbers in 1/5000s (second). A machine of this size had many operational problems, particularly with tubes burning out and circuit wiring.

In 1947, the first transistor was produced by the Bell Laboratories. These were used as switches to control the flow of electrons. They were much smaller than vacuum tubes, had fewer failures, gave off less heat, and were much cheaper to make. Computers were then assembled using transistors, but still the problem of extensive hand wiring existed. This problem led to the development of the printed circuit.

In the late 1950s, Jack Kilby of Texas Instruments and Robert Noyce of Fairchild discovered that any number of transistors along with the connections between them could be etched on a small piece of silicon (about $1/4 \times 1/4 \times 1/32$ in.). These chips, called *integrated circuits* (ICs), contained entire sections of the computer, such as a logic circuit or a memory register. These chips have been further improved so that today thousands of transistors and circuits can be crammed into the tiny silicon chip shown in Fig. 2-4. The only problem with this advanced chip was that the circuits were rigidly fixed and the chips could only do the duties for which they were designed.

In 1971, the Intel Corporation produced the microprocessor—a chip which contained the entire *central processing unit* (CPU) for a simple computer. This single chip could be programmed to do any number of tasks, from steering a spacecraft to operating a watch or controlling the new personal computers.

Thousands of bits of information can be stored on a tiny silicon chip. (*Rockwell International*)

The Role of the Computer

The first computer, developed in the early 1950s, was large and subject to breakdowns due to the many vacuum tubes it contained. As the electronics industry progressed from vacuum tubes to transistors, solid-state components, ICs, and then microprocessors, computers became smaller, more reliable, and less expensive. Today a computer no larger than a typewriter is commonplace in industry and in many homes.

The computer (Fig. 2-5) is only a tool, but it can perform many tasks with amazing speed, accuracy, and reliability. It does not have a brain and therefore cannot think for itself. The computer is only an *extension* of a person's brain and must be told exactly, in language that it understands, what it is to do. The computer can be likened to a machine tool, which is an extension of a person's muscles. The computer is fast, accurate, and stupid, while a person is slow, inefficient, and brilliant. A very useful manufacturing tool has evolved by combining the brilliance of a human with the speed and accuracy of a computer.

Although today's computers may amaze us, they have become part of our everyday life. They will continue to do so even to a much greater extent in the years to come.

We are amazed that some computers today can perform 1 million calculations per second because of the thousands of transistors and circuits jammed into the tiny chips (ICs). Computer scientists can foresee the day when 1

Fig. 2-5   The computer can perform many tasks with amazing speed, accuracy, and reliability. (*Allen-Bradley*)

billion transistors or electronic switches (with the necessary connections) will be crowded into a single chip. A single chip will have a memory large enough to store the text of 200 long novels. Advances of this type will decrease the size of computers considerably.

Some computer scientists feel that by 1990 the prototype of a thinking computer incorporating *artificial intelligence* (AI) will be introduced. They feel that the commercial product will be available about five years later. This machine will be able to recognize natural speech and written language and will be able to translate and type documents automatically. Once a verbal command is given, the computer will act, unless it does not understand the command. At this point the machine will ask questions until it is able to form its own judgments and act. It will also learn by recalling and studying its errors.

Today computers are being used in larger medical centers to catalog all known diseases with their symptoms, known cures, etc. This information is more than any doctor could remember. Doctors are now able to patch their own computer into the central computer and get an immediate and accurate diagnosis of the patient's problem, thus saving many hours and even days awaiting the results of routine tests.

Because of the computer, children will learn more at a younger age, and as may be expected, future generations will have much broader and deeper knowledge than those of past generations. It was said that in the past humanity doubled its knowledge every 25 years. Now, with the computer, the amount of knowledge is said to double every 3 years. With this greater knowledge, the human race will explore and develop new sciences and areas that are unknown to us today, much the same as the computer has done to the older people of this era.

In other areas, the computer has been and will continue to be used to forecast weather, to guide and direct planes, spaceships, missiles, and military artillery, and to monitor industrial environment.

In our everyday life, everyone is and will continue to be affected by the computer (Fig. 2-6). Department store computers list and total your purchases, at the same time keeping their inventory up to date and advising the company of people's buying habits. Thus the computer permits the company to buy more wisely. Credit bureau computers know how much every adult

Fig. 2-6   The computer has changed the way people live, work, and play. (*Hewlett Packard*)

owes, and to whom and how the debt is being repaid. School computers record students' courses, grades, and other information. Hospital and medical records are kept on anyone who has been admitted to a hospital.

Police agencies have access to a national computer which can produce the police records of any known offender. The census bureaus and tax departments of any country have the necessary information on all of its citizens on computer.

On the office front, computers have relieved accountants of the drudgery of repetitious jobs such as payroll processing. In the future, it is expected that many office workers will work from their home on a company computer. This will eliminate the necessity of traveling long distances to work and the need for baby-sitting services required by so many young working people.

In air defense control systems, the position and course of all planes from the network of radar stations are fed into the computer along with the speed and direction of each. The information is stored, and the future positions of the planes are calculated.

In the manufacturing industry the computer has contributed to the efficient manufacture of all goods. It appears that the impact of the computer will be even greater in the years to come. Computers will continue to improve productivity through *computer-aided design* (CAD), by which the design of a product can be researched, fully developed, and tested before production has started (Fig. 2-7). *Computer-assisted manufacturing* (CAM) results in less scrap

Fig. 2-7   CAD systems are invaluable to engineers who research and design products.

and more control reliability through the computer control of the machining sequence and the cutting speeds and feeds.

Robots, which are computer-controlled, are being used by industry to an increasing extent. Robots can be programmed to paint cars, weld, feed forges, load and unload machinery, assemble electric motors, and perform dangerous and boring tasks formerly done by humans.

Functions of a Computer in Numerical Control (NC)

Computers fill three major roles in NC:

1. Almost all machine control units (MCUs) built today include a computer or incorporate a computer in their operation. These control units are generally called *computer numerical controls* (CNC).

2. Most of the part programming for NC machine tools is done with off-line computer assistance.

3. There is an increasing number of machine tools which are controlled or supervised by computers which may be in a control room or even in another plant. This is more commonly known as *direct numerical control* (DNC).

The computer has also found many uses in the overall manufacturing process. It is used for such things as part design (CAD), testing, inspection, quality control, planning, inventory control, gathering of data, work scheduling, warehousing, and many other functions in manufacturing. The computer is having, and will continue to have, profound effects on manufacturing techniques in the future.

TYPE OF COMPUTERS

Most computers fall into two basic types, either analog or digital. The *analog* computer does not work directly with numbers and has been used primarily in scientific research and problem solving. Analog computers have been replaced in most cases by the digital computer.

Most computers used in industry, business, and at home are of the electronic *digital* type. It is this type that is used in numerical control work and the one explained in this book. The digital computer accepts an input of digital information in numerical form, processes that information according to pre-stored or new instructions, and develops output data (Fig. 2-8).

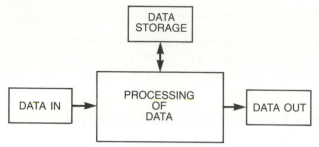

Fig. 2-8 Three functions of a computer are to accept data, process data, and output data. (*Modern Machine Shop*)

There are generally three categories of computers and computer systems: the mainframe computer, the minicomputer, and the microcomputer. Although each of these will basically perform the same tasks, some are better suited to certain applications.

The *mainframe computer* (Fig. 2-9), which can be used to do more than one job at the same time, is large and has a huge capacity. It is generally a com-

Fig. 2-9 The mainframe computer is larger and has more capacity than other computers. It can do many jobs simultaneously. (*Hewlett Packard*)

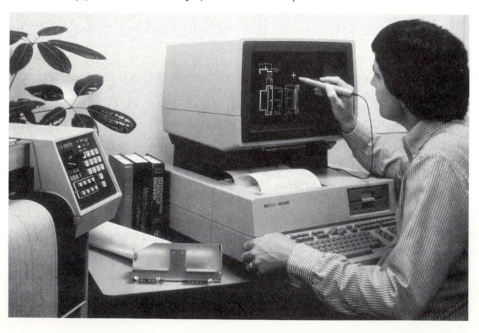

pany's main computer and the one which performs general-purpose data processing, such as NC part programming, payroll, cost accounting, inventory, and many other applications. The mainframe computer generally has a number of individual keyboard terminals connected to it, and each of these can feed information to the mainframe computer simultaneously. The CPU is designed to accept word lengths of at least 32 bits (a *bit* is a binary digit, either a 1 or a 0).

The *minicomputer* (Fig. 2-10) is generally smaller in size and capacity than the mainframe computer. This type of computer is generally of the "dedicated" type, which means that it will perform a specific task such as the:

1. Generation of part programs

2. Distribution of program data for a part of various NC machines (DNC)

3. NC of a single machine tool (CNC)

4. Management of inventory and scheduling

The minicomputer's CPU is generally designed to accept a maximum word length of 16 bits; however, some of the more recent models can accept 32 bits.

Fig. 2-10 The minicomputer is generally a dedicated type of computer and performs specific tasks. (*Hewlett Packard*)

The *microcomputer* (Fig. 2-11) generally contains one chip (a microprocessor) which contains at least the arithmetic-logic and the control-logic functions of the CPU. The microprocessor is generally designed for simple applications and must be accompanied by other electronic devices (usually on a printed circuit board) for more complex applications. The CPU is usually designed to accept 8 and 16 bits; however, there are 32-bit microprocessors available.

Computer Functions

The function of a computer is to receive coded instructions (input data) in numerical form, process this information, and produce output data which causes a machine tool to function. There are many methods used to put information into a computer. Some of the more common are punched cards, punched or perforated tape, magnetic tape, floppy disks, and specially designed sensors (Fig. 2-12). At this time, the most commonly used method to input data used in NC is the punched tape.

Output data (information coming out of the computer) may be by punched cards, punched tape, printout sheets, magnetic tape, or even directly to an-

Fig. 2-11   The microcomputer generally contains only one chip. (*Hewlett Packard*)

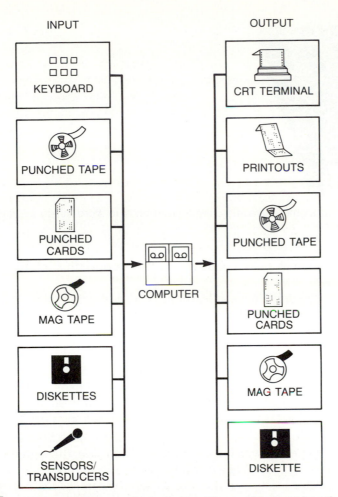

INPUT

OUTPUT

KEYBOARD

PUNCHED TAPE

PUNCHED CARDS

MAG TAPE

DISKETTES

SENSORS/ TRANSDUCERS

COMPUTER

CRT TERMINAL

PRINTOUTS

PUNCHED TAPE

PUNCHED CARDS

MAG TAPE

DISKETTE

Fig. 2-12   Several different methods are used to provide input into the computer and output out of the computer. (*Modern Machine Shop*)

other computer. Large amounts of information are generally stored on floppy-disk packs. Output data may also be sent directly to servomechanisms which activate the movements of a machine tool.

Computer Storage (Memory)

Computer storage is often referred to as "memory" to associate it with the human memory. The word *memory* may be misleading because the computer does not have the ability to think. The storage unit (memory) has the ability to

Fig. 2-13 Magnetic core memory uses tiny ferrite (iron) rings to store binary information. (*Modern Machine Shop*)

store information in a systematic way and to retrieve this information quickly and accurately.

All computers must have at least one, and preferably more, types of memory to store information for future use. These storage places are generally classified as primary or secondary storage.

1. The *primary storage* provides each binary bit with a specific storage location. It must also be able to switch each bit (0 or 1) of information ON or OFF at amazing speed. The most common storage devices are magnetic core, solid-state, and random-access memory (RAM).

   a. *Magnetic core memory*, shown in Fig. 2-13, uses tiny ferrite (iron) rings or cores which have three wires passing through them. Each binary bit (0 or 1) of a computer word (up to 32 bits) is assigned to a specific location in the core. When the core is ON (polarized in a clockwise direction), the bit represents a binary 1. When the core is OFF (polarized in a counterclockwise direction), the bit represents a binary 0.

   b. *Solid-state memory*, which is replacing magnetic core memory, uses quite a number of electronic chips, each of which can hold a large number of binary bits. *Flip-flop gates* on the circuitry of each chip provides the ON and OFF conditions necessary for handling binary numbers.

   c. *Random-access memory* (RAM) is the most important advance in computer memory because each chip has a greater capacity, and the chips are more reliable. The initial computer chips have grown from 4K (4096 bits) of memory to as high as 256K in a short period of time. A chip having 1,000,000K memory can be a reality due to the manufacturers' ability to produce higher-density memory chips. RAM allows the state of each bit (0 or 1) to be changed about 5 million times per second (once every 200 nanoseconds).

2. The *secondary storage* generally uses magnetic disks for storing information. Another form of secondary storage is by magnetic bubble memory (Fig. 2-14). This type is the one recommended for the secondary storage for NC.

   Bubble memory is a latency (resting) memory, and it is slow in comparison to RAM. While the RAM allows the state of each bit to be changed 5 million times per second, the bubble memory allows it to be changed only 25,000 times per second. However, it is more than adequate for NC functions such as storing part programs, test routines, etc.

Fig. 2-14 Bubble memory uses a rotating magnet field to move the magnetic bubbles from position to position and to represent different binary notations. (*Modern Machine Shop*)

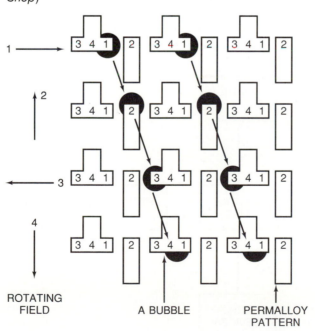

CPU

The most important functioning unit of any computer is the CPU. It contains three important sections or elements: the arithmetic-logic unit, the control unit, and the memory unit. The CPU's main job (Fig. 2-15) is to handle information as follows:

1. From the input devices to the control unit

2. Back and forth from the control unit and the arithmetic-logic unit

3. Back and forth from the control unit to the storage (memory) unit

4. From the control unit to the output devices

The arithmetic-logic unit does all the calculations and logic (comparisons or decisions) required for the program active in the computer. It handles the basic operations of adding, subtracting, multiplying, and counting.

The memory unit provides short-term or temporary storage of data (information) being processed at any particular time. It is also closely tied in with the main memory of the computer, and therefore stored information may be retrieved very quickly. The memory section of the CPU contains a memory

Fig. 2-15   The main functions of the CPU.

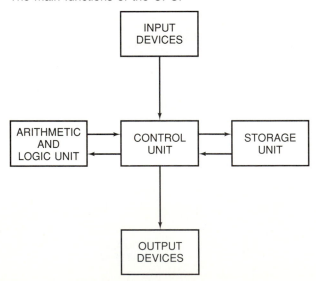

address register whose function is to store data in a specific place and to retrieve it from this place when needed.

Since we are now in the Computer Age, there will be more and more demand for people who are knowledgeable about and capable of working with computers in all aspects of manufacturing. It is important that everyone who expects to be associated with some form of manufacturing have a working knowledge of computers. The computer is a useful tool which can help a person to produce work faster, with fewer errors, and at a lower cost. The future is bright, exciting, and challenging; be sure you are prepared for it.

REVIEW QUESTIONS

History of the Computer

1. Name three methods of calculation used by primitive people.

2. What was the first computer ever developed?

3. For what purpose were the first punched cards used in the United States?

4. How many transistors and circuits can be found on a silicon chip?

5. Name three advantages of using a computer to perform tasks.

6. How are computers used in medical centers?

7. How are computers affecting the manufacturing industry?

8. List eight uses for a computer in the overall manufacturing process.

Computer Functions

9. What is the most common type of computer used in industry?

10. Name three functions that a digital computer performs in NC.

Types of Computers

11. Name three categories of computer and computer systems.

12. What methods may be used to put information into a computer?

13. Define *input data* and *output data*.

14. Briefly explain primary memory (storage).

15. Name three types of primary memory.

16. How does bubble memory compare to RAM in regards to speed of bit changes?

Central Processing Unit

17. List the three important elements of a CPU.

18. Explain the function of each of the three elements in a CPU.

CHAPTER

THREE

Input Media

Numerical control (NC) operates machine tools by sending a series of coded instructions consisting of alphabet letters, numbers, and symbols in language that the machine control unit (MCU) can understand (Fig. 3-1). These coded instructions are turned into pulses of electrical current or other output signals which operate the motors and servomechanisms of the machine tool. The coded instructions or commands are listed in a logical sequence to have a machine tool perform a specific task or a series of tasks in order to produce a finished product.

After completing this chapter you should be able to:

1. List the advantages and disadvantages of various NC input media

2. Understand the standard NC tape coding system

3. Know the sequence to follow in order to produce an NC tape

TYPES OF INPUT MEDIA

As NC developed over the years, a number of different media were used to convey the information from a drawing to the MCU. The most common types of input were manual, punched card, magnetic tape, punched tape (Fig. 3-1), and diskette (Fig. 3-2).

Manual Data Input (MDI). Data that is input manually can be used to program the control system by setting the dials, switches, push buttons, etc. Although this method is not often used because it is slow and subject to operator inaccuracies, most NC machine tools can be programmed manually, especially for setup purposes.

Punched Cards. Some NC systems used 80- or 90-column punched cards, which carried a great deal of information. The card reader could decode as many as 120 cards per minute. The disadvantage of punch cards was that under typical shop conditions, the cards became bent and would jam in the card reader. Also, the sequence of the cards could be altered if extreme care was not used in handling them.

Magnetic Tape. In the late 1950s, some NC machine control units (MCUs) used 1-in. (25-mm) magnetic tape to store data. This tape was similar to what was used to record music and conversation, but was of a higher quality. It

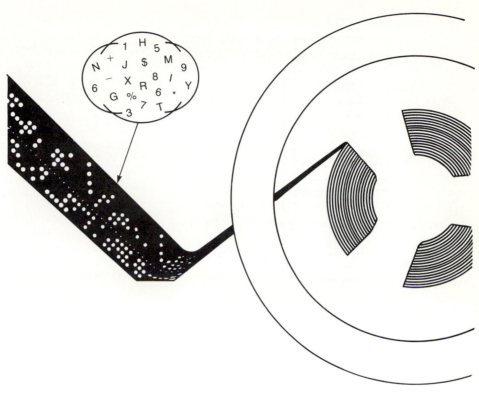

Fig. 3-1 NC operates machine tools by sending a series of coded instructions (alphabet, numbers, and symbols) to the MCU. (*Cincinnati Milacron, Inc.*)

was not used for any length of time because interference from nearby electrical equipment such as transformers and other shop equipment had a tendency to erase or scramble some of the information on the tape.

Today magnetic tape has better shielding from outside electrical interference, and the ¼-in. (6-mm) tape in a tape cassette is again being used for some NC applications.

Punched Tape. Because punched tape is the most common form of input medium, it has been adopted as the standard for the NC industry. The 1-in. (25-mm) eight-channel tape (Fig. 3-3) using the binary-coded decimal (BCD) system has been selected as the standard by the Electronic Industries Association (EIA). Holes are punched in the channels, which run lengthwise, and the placement of the holes, in horizontal rows, determines the operating instructions for the machine tool. The tape reader on the MCU decodes the pattern of holes, generally by a photoelectric reader, and converts these into electrical

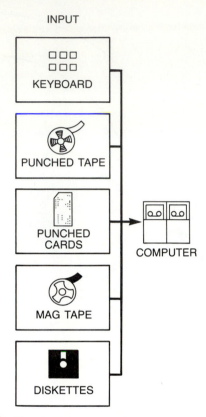

INPUT

KEYBOARD

PUNCHED TAPE

PUNCHED CARDS

COMPUTER

MAG TAPE

DISKETTES

Fig. 3-2 Punched tape, punched cards, magnetic tape, diskette, and manual data input have been used to input data into an MCU. (*Modern Machine Shop*)

pulses which operate the motors and the servomechanisms of the machine tool.

Punched tapes are made from several different types of materials, including paper, Mylar (a type of plastic), and foil.

1. *Paper:* Tapes made of paper are available in a variety of colors and may be purchased oiled or unoiled. The oiled tapes are recommended because they help to lubricate the punches in the tape-preparation equipment used to punch the holes in the tape. Paper tape is inexpensive, but it must be used with care since it tears easily. Paper tapes are generally used for file purposes.

2. *Mylar:* Tapes made of Mylar, laminated between two strips of paper, have proved sturdy and almost impossible to tear. Although it is more expensive than the plain paper tape, it is highly recommended

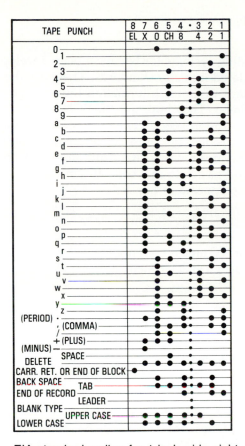

Fig. 3-3 EIA standard coding for 1-inch-wide eight-channel tape.

for shop use because of its strength. Aluminized Mylar tapes are recommended for light-source tape readers. However, they are slightly more expensive than the paper Mylar tape and also create considerable wear on the punches used to produce the tape.

3. *Foil:* Foil tapes are sturdy but not generally recommended because they are very hard on the punches of the tape-preparation equipment.

Diskette. The diskette, or floppy disk, is very much like a small flexible phonograph record which has channels or tracks to store magnetic bits of information. One diskette can store as much information as 2000 to 8000 ft of punched tape. The bits of information on the diskette can be decoded by a reading head on the MCU. The major advantage of diskettes is that it is much faster to retrieve information from this medium than from any other.

TAPE-CODING SYSTEMS

Two well-known tape-coding systems for NC have been developed over the years and are now considered industry standards. These are the EIA system and the American Standard Code for Information Interchange (ASCII) system. The two are very similar. They both use the BCD system for numerical data and both use the 1-in. (25-mm) eight-channel tape. There is a demand for both types of coding, and this does not present a problem to control manufacturers or the programmer, since today's MCUs and tape-punching units recognize the type of code and switch to that code automatically. A comparison of the two coding systems appears in Fig. 3-4.

EIA Coding System

The EIA coding system for NC tape is commonly used to control NC machine tools and equipment. It uses a BCD system on 1-in. (25-mm) eight-channel tape for numerical data.

1. *Five of the channels (1 to 4) are assigned the values of 0, 1, 2, 4, and 8,* so that any numerical quantity from 0 to 9 can be inserted in one horizontal line of the tape (Fig. 3-4).
   a. To indicate the digit 7, holes would be punched in channels 4, 2, and 1.
   b. The digit 3 would require holes punched in channels 2 and 1.
   Each numerical digit, letter, or symbol code has its own combination of holes in a *single row* on the tape.

2. *The fifth channel* (indicated by CH) is the odd-parity bit and is used as a safety device to reduce the chance of error when preparing tape. The EIA coding system states that there must be an odd number of holes in every row of information. This is called the *odd-parity system*. Assume that the digit 6 is required. Holes would have to be punched in the same row in columns 3 and 2 (digits 4 and 2). Because this would be 2 holes, an even number, the tape preparation equipment automatically punches a hole in channel 5 (Fig. 3-5).

3. *The sixth channel* (indicated by 0) is a special binary code and is assigned to lowercase letters of the alphabet or when a 0 is required (Fig. 3-6).

4. *The seventh channel* (X) indicates the minus sign (−) and is also assigned to lowercase letters of the alphabet.

EIA STANDARD RS 244 KEYBOARD SYMBOLS	ALTERNATE KEYBOARD SYMBOLS	CODE ON TAPE — CHANNEL NUMBERS 1	2	3	•	4	5	6	7	8	SYSTEM FUNCTION
TAPE FEED	SPACE. BUZZ.FEED				•						LEADER
	RWST. %. $	●	●		•	●	●		●		RWS (REWIND STOP)
TAB			●	●	•	●	●	●			TAB
+					•		●	●	●		+ (OPTIONAL)
−					•				●		−
1		●			•						1
2			●		•						2
3		●	●		•		●				3
4				●	•						4
5		●		●	•		●				5
6			●	●	•		●				6
7		●	●	●	•						7
8					•	●					8
9		●			•	●	●				9
0					•			●			0
CAR. RET. OR EOB					•					●	EOB (END OF BLOCK)
DELETE	TAPE FEED	●	●	●	•	●	●	●	●		DELETE

A

ASCII STANDARD SYMBOLS	CODE ON TAPE — CHANNEL NUMBERS 1	2	3	•	4	5	6	7	8	SYSTEM FUNCTION
NUL				•						LEADER
EOT			●	•					●	RWS (REWIND STOP)
TAB	●			•	●					TAB
+	●	●		•	●		●			+ (OPTIONAL)
−	●		●	•	●		●			−
1	●			•		●	●		●	1
2		●		•		●	●		●	2
3	●	●		•		●	●			3
4			●	•		●	●		●	4
5	●		●	•		●	●			5
6		●	●	•		●	●			6
7	●	●	●	•		●	●		●	7
8				•	●	●	●		●	8
9	●			•	●	●	●			9
0				•		●	●			0
CR/EOB	●		●	•	●				●	EOB (END OF BLOCK)
DELETE	●	●	●	•	●	●	●	●	●	DELETE

B

Fig. 3-4 A comparison of the (A) EIA and (B) ASCII coding systems for NC tape. (The EIA system consists of 63 different chapters including numbers, letters, and symbols; ASCII has 128 characters.) (*The Superior Electric Company*)

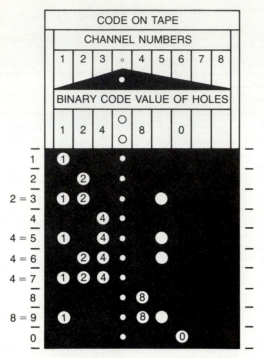

The EIA binary number code-on-tape format. (*The Superior Electric Company*)

The channels used in NC to indicate the letters of the alphabet with the BCD system.

Letter coding (binary system)		
Punch holes in channels 6 and 7 plus numerical value	Punch hole in channel 7 plus numerical value	Punch hole in channel 6 plus numerical value
1 = a	1 = j	
2 = b	2 = k	2 = s
3 = c	3 = l	3 = t
4 = d	4 = m	4 = u
5 = e	5 = n	5 = v
6 = f	6 = o	6 = w
7 = g	7 = p	7 = x
8 = h	8 = q	8 = y
9 = i	9 = r	9 = z

5. *The eighth channel* (EL) indicates the end of a block of information. It is punched at the beginning and end of most NC punched tapes and is never combined with any other holes.

ASCII Coding System

The ASCII system was developed by various committees working with the United States Standard Institute (now the American National Standards Institute). The goal was to produce one coding system which would give maximum capacity to the communications industry. Using this coding system, telephone and telegraph companies, computer companies, governments, and others have an international standard for all information processing and communications systems. This has resulted in making the ASCII system the "universal" perforated tape coding system. However, NC has been and still is heavily committed to the original EIA (RS-244) code. The ASCII subset used for NC tape is now the EIA (R-358) standard. See Fig. 3-4 for a comparison of the EIA and ASCII tape-coding formats.

Since both tape-coding systems are used and the equipment is capable of handling either, it is important to understand the differences found in each system.

Similarities

- Both EIA and ASCII use the 1-in. (25-mm) eight-channel tape.
- The codes for digits 0 to 9 are the same for both systems.
- Both systems use the BCD for numerical data.

Differences

- ASCII provides coding for both uppercase and lowercase letters of the alphabet. EIA provides coding only for lowercase letters.
- ASCII uses an even-parity check (channel 8), so there must always be an even number of holes in each row. EIA uses an odd-parity check.
- In EIA holes are punched in two additional channels (5 and 6) to identify the numbers and certain symbols.
- In EIA a hole is punched in channel 7 for all alpha characters.

Letter Coding

The EIA binary-coded decimal system uses number codes (1 to 9) in combination with channel 7 (X channel) and/or channel 6 (0 channel) for letters of the alphabet. This coding for the alphabet (letters) is also subject to the odd-parity check in channel 5 if the number of holes punched is even (Fig. 3-7).

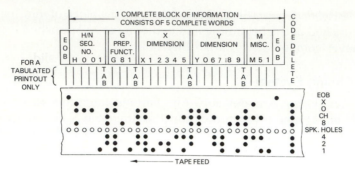

Fig. 3-7 A complete block of information on punched tape consists of five words.

TAPE FORMAT

In NC, the arrangement and sequence in which coded information appears on the punched tape is referred to as *tape format*. The coded information appears as words (a series of holes in horizontal rows), as shown in Fig. 3-7. Five complete words or pieces of information are contained in a block (one instruction). If five complete words are not included in each block, the tape reader will not recognize the information on the tape and will not activate the control unit. The information contained on the section of punched tape in Fig. 3-7 is as follows:

1. A hole is punched in channel 8, which represents the end or the beginning of a line or block of information.

2. The *first word* of the block represents the number of the operation. H001 represents the first operation on the tape. The letter H is recorded by punching holes in channels 4, 6, and 7 (Fig. 3-7).
   - The *two 0s* are recorded by punching channel 6 in two successive rows.
   - The *1* is recorded by punching channel 1, which has a numerical value of 1.

3. The *TAB* is used to separate each word into a block of information.

4. The *second word* in the block represents the type of operation to be performed. G81 is a drill cycle on Cincinnati Milacron NC systems.
   - The *letter* G is recorded by punching channels 1, 2, 3, 6, and 7.
   - The *8* is recorded by punching channel 4, which has a numerical value of 8.

- The *1* is recorded by punching channel 1, which has a numerical value of 1.

5. The *third word* represents the distance the table slide must move along the X axis. The information contained in the third word of Fig. 3-7 is X12345. Because standard tapes on closed-loop systems program all dimensions to ten-thousandths of an inch, the machine will move 1.2345 in. along the X axis.

6. The *fourth word* represents the distance the table slide must move along the Y axis. In Fig. 3-7, Y06789 represents a table movement of 0.6789 in. along the Y axis.

7. The *fifth word* represents a miscellaneous machining function. M03 would start the machine spindle revolving in a clockwise direction.

Tape Format Types

There have been a number of different formats used on NC punched tape over the years. These include fixed sequential, TAB ignore, TAB sequential, word address, and interchangeable, or compatible format. The details of each format will be explained using a five-word block of information consisting of a sequence number, X-axis dimension, and Y-axis dimension to drill the three holes through the workpiece shown in Fig. 3-8.

Fixed Sequential.   The control systems on early NC machine tools used only numerical data, in a very rigid order, with the machine control codes appearing in every block of information. This format had three disadvantages:

1. Printout copies were very difficult to read and check because all the numbers appeared as one long word.

2. Because codes were repeated in every block, the programmer took longer to program the information and, naturally, the NC punched tape was longer and subject to more error.

3. There was no word letter address used to identify individual words in the program.

TAB Ignore.   This programming format used TAB codes between each word to make it easier to read the tape; also, a much neater printout was produced (Fig. 3-8). The TAB codes are not recognized by the MCU, and therefore no action occurs as a result of their use. One disadvantage of this format is that it

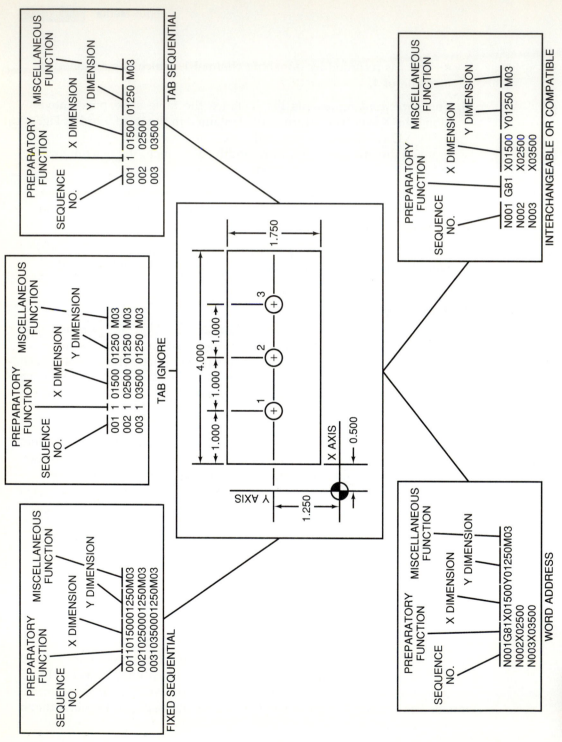

Fig. 3-8   Various types of punched tape formats.

TAB SEQUENTIAL

SEQUENCE NO.	PREPARATORY FUNCTION	X DIMENSION	Y DIMENSION	MISCELLANEOUS FUNCTION
001	1	01500	01250	M03
002		02500		
003		03500		

INTERCHANGEABLE OR COMPATIBLE

SEQUENCE NO.	PREPARATORY FUNCTION	X DIMENSION	Y DIMENSION	MISCELLANEOUS FUNCTION
N001	G81	X01500	Y01250	M03
N002		X02500		
N003		X03500		

TAB IGNORE

SEQUENCE NO.	PREPARATORY FUNCTION	X DIMENSION	Y DIMENSION	MISCELLANEOUS FUNCTION
001	1	01500	01250	M03
002	1	02500	01250	M03
003	1	03500	01250	M03

FIXED SEQUENTIAL

SEQUENCE NO.	PREPARATORY FUNCTION	X DIMENSION	Y DIMENSION	MISCELLANEOUS FUNCTION
001	1	01500	001250	M03
002	1	02500	001250	M03
003	1	03500	001250	M03

WORD ADDRESS

SEQUENCE NO.	PREPARATORY FUNCTION	X DIMENSION	Y DIMENSION	MISCELLANEOUS FUNCTION
N001	G81	X01500	Y01250	M03
N002		X02500		
N003		X03500		

still contains information which is repeated in every block and does not have codes which would make it more efficient.

TAB Sequential.   TAB codes are used in this format to separate words and also to replace similar words that are repeated in following blocks. This helped to reduce the length of time it took programmers and typists to prepare a program.

Word Address.   This format, which uses the letter address to identify each separate word, has been standardized by the EIA. The block format is much easier to program because alphabetical codes have been assigned to each coordinate or function word. This reduces the amount of data on the tape and makes it easier to identify words. Only the codes that change from one block to another need to be programmed. Since there are no TAB codes, the tape is difficult to read and check.

Interchangeable (Compatible).   This format, which is basically the same as the word address format, provides for the addition of TABs which make the tape neater and easier to read. The length of a block is variable, and words can be interchanged within a block. Interchangeable format meets EIA standards and is probably the easiest and most flexible format in use today.

TAPE PREPARATION

An NC system is an obedient servant that will faithfully follow the instructions it receives, but it cannot think for itself and cannot overcome errors in the tape program. NCs may use either absolute or incremental positioning. With absolute positioning, each machining location is given in relation to a zero point, or origin. When the incremental positioning is being programmed, each machining location is given from the last position.

The stages in preparing an NC punched tape begin with the part drawing, and progress through preparing a manuscript, writing the program, punching the tape, and finishing with the final tape (input media) ready for use in an NC machine (Fig. 3-9). Three documents will contain information regarding the part, and the fourth is the punched tape (Fig. 3-10).

Part Drawing

The part drawing or print, prepared by the engineering department, must contain all the specifications of the part, such as the shape, sizes, dimensions, and the type of operations which must be performed. In Fig. 3-11, six holes of

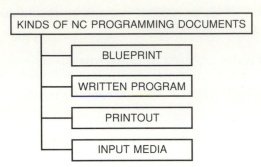

Fig. 3-9 Four essential documents are required to produce a punched tape which the machine can use to produce a finished part. (*The Superior Electric Company*)

⅜-in. diameter must be drilled in a 3⅝- × 6-in. plate. It is assumed that the plate is already to size; only the holes must be located in the proper location and drilled.

Manuscript

An NC programmer must take the information contained on the part drawing (print) and put it into a language that the MCU will understand. This is done by taking the information from the part drawing and preparing a program manuscript which lists the steps and sequences required to machine the part (Fig. 3-12).

The *programmer* plays a very critical part in whether a workpiece will be produced to exact sizes. Since an NC machine does exactly what it is programmed to do, it must be programmed accurately. Before a tape program is

Fig. 3-10 The three NC documents which contain details of the machining required on a part. (*The Superior Electric Company*)

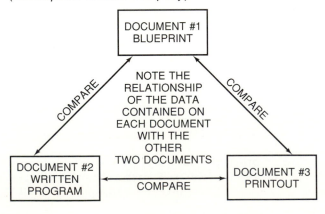

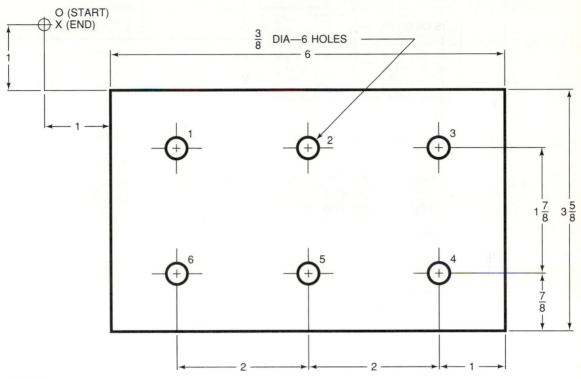

Fig. 3-11 The drawing should contain all the information necessary to produce a part to exact sizes. (*The Superior Electric Company*)

prepared, it is necessary to know whether the program is being prepared for absolute or incremental positioning. Since the role of the programmer is so essential for the successful operation of any NC machine tool, it is desirable to review the qualities a person should have to become a successful programmer (Fig. 3-13).

In many cases, it is advisable to make a programmer's sketch of the part showing the sequence of operations and dimensioning it to suit the programming mode, as in Fig. 3-14. The following steps are required to produce a manuscript (program).

1. Determine the fixture most suitable to locate and hold the workpiece on the machine table.

2. Select the zero point, or origin, to allow the workpiece and machine tool to be aligned.

PREPARED BY *LGH*	PART NAME		PART NO.	OPER. NO.
DATE 3-27-67		*PLATE*	*BG 75507*	—
CK'D BY *REC*				
DATE 3-28-67	REMARKS:			
SHEET / OF /	*RUN PROGRAM TWICE, ONCE WITH*			
DEPT *16*	*CENTER DRILL, ONCE WITH 3/8 DRILL*			
TAPE NO *127*	*SET FEED AT HI, TOOL AT AUTO, BACKLASH AT #2*			

SEQ NO	TAB OR EOB	OR	"X" INCREMENT	TAB OR EOB	OR	"Y" INCREMENT	TAB OR EOB	"M" FUNCT	EOB	INSTRUCTIONS
	EOB									
RWS	*EOB*									*CHANGE TOOL, LOAD, START*
1	*TAB*		*2000*	*TAB*	*—*	*1875*	*EOB*			
2	*TAB*		*2000*	*EOB*						
3	*TAB*		*2000*	*EOB*						
4	*TAB*			*TAB*	*—*	*1875*	*EOB*			
5	*TAB*	*—*	*2000*	*EOB*						
6	*TAB*	*—*	*2000*	*EOB*						
7	*TAB*	*—*	*2000*	*TAB*		*3750*	*TAB*	*02*	*EOB*	

Fig. 3-12 The manuscript lists the sequence of operations, tools required, and the various sizes of a part. (*The Superior Electric Company*)

This point may be at the corner of the part, a point on the holding fixture, or at any point off the job.

3. Select the first tool change point where the cutting tools can be changed and workpieces loaded and unloaded.

This may be at the zero point or at a second point which would allow parts to be loaded or unloaded easily.

4. Determine the sequence of machining operations necessary to machine the part accurately and in the least amount of time.

5. Record the sequence of operations on the manuscript (program) form. Include operator's instructions such as tool changes or speeds.

NOTE: The manuscript layout must be the same as the tape format to be used.

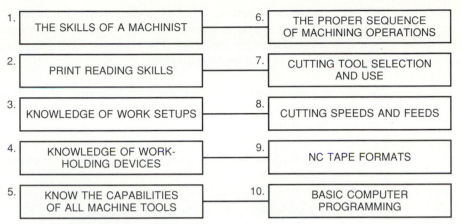

1. THE SKILLS OF A MACHINIST	6. THE PROPER SEQUENCE OF MACHINING OPERATIONS
2. PRINT READING SKILLS	7. CUTTING TOOL SELECTION AND USE
3. KNOWLEDGE OF WORK SETUPS	8. CUTTING SPEEDS AND FEEDS
4. KNOWLEDGE OF WORK-HOLDING DEVICES	9. NC TAPE FORMATS
5. KNOW THE CAPABILITIES OF ALL MACHINE TOOLS	10. BASIC COMPUTER PROGRAMMING

Fig. 3-13   Desirable qualities of a programmer.

For an example of the steps a programmer must follow, see Fig. 3-11. In this example, point-to-point positioning in the absolute mode and the interchangeable (compatible) format will be used to prepare the tape to drill the six holes. For simplicity in this manuscript, let us assume that the holes will not be center-drilled before the ⅜-in. holes are drilled, and no tool changes are required.

Each horizontal line on the manuscript (Fig. 3-12) contains one block of information for one positioning movement along with the miscellaneous functions (tool changes, speeds, coolant, etc.) that are required. Let us examine each piece of information in the block of information in Fig. 3-12.

N010—The Sequence Number.   Each block of information is assigned a sequence number, and this is the first item entered on the manuscript. The sequence number identifies each block of information and allows the information to be recalled at any time for reference or revision. The sequence numbers are generally listed in progressive order by 10s, and they can be preceded by the letters H, N, or O.

Fig. 3-14   A programmer's sketch dimensioned for incremental positioning and listing the machining sequence. (*The Superior Electric Company*)

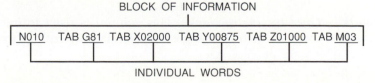

BLOCK OF INFORMATION

N010   TAB G81   TAB X02000   TAB Y00875   TAB Z01000   TAB M03

INDIVIDUAL WORDS

G81—Preparatory Function. The G code refers to functions such as positioning systems, the direction of spindle revolvement, canned and fixed cycles, and programming mode. In this example, G81 refers to a drilling operation which is one of the fixed cycles.

Numbering Operations. Line numbers such as 010, 020, 030, and so forth, allow new information to be inserted between any operation without the necessity of renumbering all of them.

TAB. The TAB code should be entered before and after entering a word to make the tape easier to read.

X02000—X Increment. The amount of movement along the X axis is indicated in this location. In absolute positioning, all distance (movements) are taken from the zero point, or origin. A movement to the right of the zero point is a plus movement; movement to the left is a minus movement. Plus movements do not have to be entered with the plus symbol (+), but minus movements must be indicated with the minus symbol (−) on the manuscript. In this example, the cutting tool must be moved 2.000 in. to the right of the zero point.

Y00875—Y Increment. The amount of movement along the Y axis is indicated in this location. A movement up or away from the zero point would be a plus movement. A movement down or toward the zero point would be a minus movement. In this example, the cutting tool must be moved 0.875 in. up from the zero point.

Z01000—Z Increment. The amount of movement along the Z axis is indicated in this location. A movement toward the Z zero point or gage height would be a minus movement. A plus movement would be away from the Z zero point. In this case, the spindle of the machine holding the drill would move 1.000 in. above the Z zero point.

M03—Miscellaneous Function. Any miscellaneous function codes such as tool change, tool description and size, rapid feed rate, rewind codes, etc., are indicated in this location. In this case, the M03 code will cause the cutting tool to revolve in a clockwise direction.

Remarks—Operator Instructions. This column is used to inform the machine tool operator of tool changes, types, sizes of cutting tools, or other actions which are necessary.

TAPE-PRODUCING PROCESS

After the manuscript has been completed and checked for accuracy, the punched NC tape can be prepared in a number of ways.

Manual Preparation

At one time, the perforated (punched) tape was produced on tape-preparation equipment similar to a typewriter (Fig. 3-15). The programming clerk would key in the information from a manuscript on the tape typewriter, which would activate the punches in the tape writer to punch holes in the proper location on the tape. The program was punched on tape in much the same manner as a teletype machine processes data. When the complete manuscript (program) was typed, all the data required to produce a part was checked for accuracy. This tape could be used at any time it was necessary to produce parts on the NC machine.

Although some of the older tape typewriters may still be in use, most tape is prepared on a computer—by keying in the information from a manuscript. A programming clerk keys in the information from the manuscript starting at the top of the sheet (Fig. 3-16). Each line of the manuscript represents one block of information on the tape. Once the entire program has been entered,

Fig. 3-15 A manual tape-punching typewriter which converts the data from the manuscript to punched tape ready for use in an NC machine. (*Modern Machine Shop*)

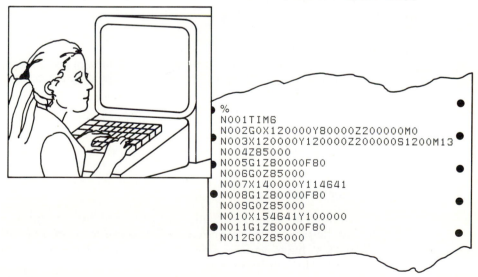

```
%
N001TIM6
N002G0X120000Y80000Z200000M0
N003X120000Y120000Z200000S1200M13
N004Z85000
N005G1Z80000F80
N006G0Z85000
N007X140000Y114641
N008G1Z80000F80
N009G0Z85000
N010X154641Y100000
N011G1Z80000F80
N012G0Z85000
```

Fig. 3-16 A programming clerk keys information from the manuscript into a computer which can then produce a punched tape or a printout of the program or plot the job on the screen. (*Facit, Inc.*)

it is checked and corrected if necessary, and then the computer activates the tape-punching unit to produce the tape. While the program is in the computer, it can also activate the printer to produce a printout record of the program.

Computer Preparation

NC programs and tapes can often be prepared by computer systems with very little help from the programmer. For example, a circuit board containing 2000 to 3000 holes can be designed using computer-aided design (CAD) equipment. This equipment can produce the prints (drawings) and artwork necessary for the manufacture of circuit boards and also make the punched tape to run the drilling machines which produce the holes. With this system, no manual programming is required, since the locations for each hole are known from the generated artwork, and from that the computer prepares a program similar to the one produced by a programming clerk on tape preparation equipment.

Editing Programs

Regardless of how a program has been prepared (manually or computer-assisted), the final criterion of whether it is good or not depends on its successful execution. If it produces scrap work, damages the machine tool, breaks a cutter, jams, etc., it is quite obvious that something in the program is not correct. Most new programs have some "bugs" in them, and it is wise to discover them before the work is scrapped or the machine damaged.

The programmer must review the manuscript to check for errors, wasted machine motion, and repetition of unnecessary information. Wherever possible, the manuscript and the computer printout should be compared to catch any obvious errors (Fig. 3-17). The printout should also be used to check the accuracy of data transmitted over telephone lines to the NC machine tool at another location.

Regardless of how carefully all sources are checked, some error will always slip through because of the human factor. There are several editing systems on the market which have been specifically developed to check the accuracy of a program before it is used on a machine.

Fig. 3-17 The manuscript, printout, punched tape, and transmitted data should be checked for errors before the program is used to control a machine tool. (*Hewlett Packard*)

1. *CRT (Cathode-Ray Tube) Screens or Plotting Units*

   Many of the newer programming systems use a minicomputer or microcomputer with either a CRT screen or a plotting device (Fig. 3-18). Visual images of the part can be created which will quickly show errors such as incorrect contours or a misplaced decimal in a dimension. A dimension of 2.5 in. could easily be either 0.25 in. or 25 in., depending on the placement of the decimal, and this difference would become very obvious when viewed on the screen. The plotter device can be used to generate the cutter path to see that it produces the correct shape of the part.

2. *Shop-Floor Programming Systems*

   Viewing screens and keyboards are now standard equipment for many computer numerical control (CNC) shop-floor programming units. The visual part can be seen on the screen right at the machine tool, and any corrections or revisions can be made to the part before the machine tool is run.

3. *Portable Editing Units*

   Portable editing units consisting of tape readers, punches, data readout, CRT screens, etc., have been developed. They can be taken directly to the machine tool where program changes may be necessary or where a program must be checked. This is especially useful when it is impossible or not convenient to take the program back to the programming area for corrections or revisions.

Fig. 3-18 CRT screens and plotting devices can be used to check the accuracy of NC punched tape. (*Hewlett Packard*)

The newer programming systems and controls have made part programming, editing, and revising much easier than the early complicated NC systems. Regardless of how good the equipment is, there is always the possibility of an error in the program because of its length and the human factor involved. Until a foolproof system is devised, some program editing and revising will always be necessary.

REVIEW QUESTIONS

1. Briefly explain how numerical control operates.

Types of Input Media

2. Name five types of input media which have been used in NC.

3. Describe the standard input medium which has been selected by the EIA.

4. Name three types of material used for NC punched tape.

5. What are the advantages of using diskettes for storing NC information?

Tape-Coding Systems

6. Name the two most common types of tape-coding systems used in the NC industry.

7. What channels on the EIA tape are assigned to the numerical values of 1, 2, 4, and 8?

8. What is the purpose of the fifth channel on the EIA tape?

9. State the purpose of the sixth, seventh, and eighth channels on EIA tape.

10. Why was the ASCII tape developed?

11. Compare the similarities and differences of EIA and ASCII numerical control tape.

Tape Format

12. What does one block of information consist of?

13. Why is it important that five words appear in each block?

14. Explain what each of the five words in a block represent.

15. Name five types of formats used on NC tapes.

16. Compare the fixed sequential and the TAB ignore format.

17. Compare the word address and the interchangeable format and state the advantages these have over previous formats.

Tape Preparation

18. Explain absolute and incremental positioning.

19. Name three documents which contain information regarding the tape program.

20. What role does the part drawing play for NC programming?

21. What is the purpose of a manuscript?

22. Why is the programmer so important to NC work?

23. List five of the most important qualifications that a good programmer must have.

24. Briefly list the five steps required to produce a manuscript (program).

25. Briefly explain sequence number, preparatory function, and miscellaneous function.

26. Explain X and Y increment and the direction of movement for each when there is a plus or minus movement.

Tape-Producing Process

27. What is tape-preparation equipment, and how does it produce NC punched tape?

28. How can computers be used to produce NC tape and greatly reduce the amount of time a programmer spends?

29. Why is it important that NC tape programs be checked before they are used on a machine?

30. What should the programmer look for when editing a program?

31. How can a CRT screen or a plotting unit help to check the accuracy of an NC program.?

32. State the advantages of shop-floor programming systems and portable editing units.

CHAPTER

FOUR

How Numerical Control Operates Machine Tools

Numerical control (NC) operates the machine tool in more or less the same way as an operator would, but it is done automatically. NC offers almost unbelievable savings in production costs, part accuracy, and many other benefits. The following steps summarize how NC works:

1. Numerical data may be fed into the system by punched tape, floppy disk, or magnetic tape, or directly from a computer

2. A translating unit reads the data and changes it into an electrical form that the machine tool can understand

3. A memory system stores the data until it is needed

4. Servo units (transducers) on the machine tool convert the data into actual machine movements

5. A gaging device measures the machine movements to determine if the servo units have given the correct commands

6. A feedback unit feeds information back from the gaging device for comparison so that the machine moves to the correct location

Once the punched tape for a particular part has been checked for accuracy, it is ready to use. The tape contains, in binary form, complete information for moving the machine table into each machining location. No jigs are required; a simple holding fixture is all that is needed to locate and hold the part. The punched tape accurately guides the machine table to the correct location, stops there, revolves the cutting tool, sets speeds and feeds, and performs all the necessary operations in order to produce the finished part. NC is now used on all types of machine tools, electrical discharge machines, welding machines, and inspection systems, and for most manufacturing and assembly processes.

Objectives

After completing
this chapter,
you should be
able to:

1. Understand the purpose of the tape reader

2. List the advantages and disadvantages of open- and closed-loop systems

3. Know the function of the machine control unit (MCU) and the central processing unit (CPU)

Starting with the edited punched tape for a specific part, many elements are involved, from the control system to the machine tool, to decode and process this information and activate the machine tool in order to produce a finished part. The key elements are the tape reader, NC control systems, servomechanisms, MCU, CPU, etc. (Fig. 4-1). Since all these elements play a very important part in the NC of machine tools, each will be covered in detail.

NC PERFORMANCE

NC has made great advances since it was first introduced in the mid-1950s as a means of guiding machine tools through various motions automatically, without human assistance. The early machines were capable only of point-to-point positioning (straight-line motions), were very costly, and required highly skilled technicians and mathematicians to produce the tape programs.

Great advances in NC came as the result of new technology in the electronics industry. The development of transistors, solid-state circuitry, integrated circuits (ICs), and the computer chip have made it possible to program machine tools to perform tasks undreamed of as recently as a decade ago. Not only have the machine tools and controls been dramatically improved, but the cost has been continually dropping. NC machines are now within the financial reach of small manufacturing shops and educational institutions. Their wide acceptance throughout the world has been a result of their accuracy, reliability, repeatability, and productivity (Fig. 4-2).

Accuracy

NC machine tools would never have been as well accepted by industry if they were not capable of machining to very close tolerances. At the time NC was being developed, industry was looking for a way to improve production rates and achieve greater accuracy on their products. A skilled machinist is capable of working to close tolerances, such as ±0.001 in. (0.025 mm) or even less on most machine tools. It has taken the machinist many years of experience to develop this skill, but this person may not be capable of working to this accuracy every time. Human error will always result in some mistakes, which means that the product may have to be scrapped.

Modern NC machine tools are capable of producing workpieces which are accurate to within a tolerance of 0.0001 to 0.0002 in. (0.0025 to 0.0050 mm). The machine tools have been built better, and the electronic control systems ensure that parts within the tolerance allowed by the engineering drawing will be maintained. The accuracy, formerly dependent on the machinist's

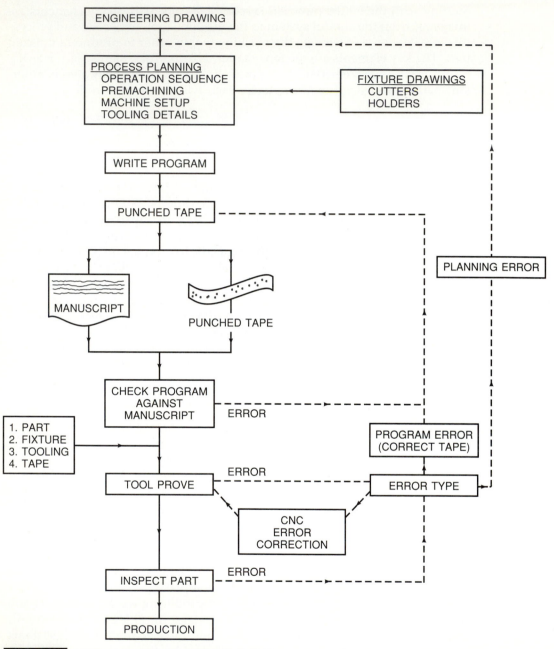

ENGINEERING DRAWING

PROCESS PLANNING
  OPERATION SEQUENCE
  PREMACHINING
  MACHINE SETUP
  TOOLING DETAILS

FIXTURE DRAWINGS
  CUTTERS
  HOLDERS

WRITE PROGRAM

PUNCHED TAPE

PLANNING ERROR

MANUSCRIPT

PUNCHED TAPE

CHECK PROGRAM
AGAINST
MANUSCRIPT
ERROR

1. PART
2. FIXTURE
3. TOOLING
4. TAPE

PROGRAM ERROR
(CORRECT TAPE)

TOOL PROVE
ERROR

ERROR TYPE

CNC
ERROR
CORRECTION

INSPECT PART
ERROR

PRODUCTION

Fig. 4-1   NC manual programming flowchart.

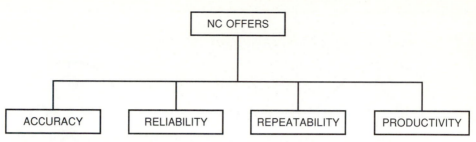

Fig. 4-2 NC has been widely accepted because of the many advantages it has to offer industry.

skill, is now being achieved by NC, with reliable control systems and better machine tool construction.

Reliability

The performance of NC machine tools and their control systems had to be at least as reliable as their skilled help (machinists, toolmakers, and diemakers, etc.) for industry to accept this new machining concept. Since consumers throughout the world were demanding better and more reliable products, there was a great need for equipment that could machine to closer tolerances and be counted on to repeat this time and time again.

Machine tools were greatly improved when all parts and components were made to closer tolerances. Improvements in machine slides, bearings, ball screws, and machine tables all helped to make sturdier and more accurate machines. New cutting tools and toolholders were developed which matched the accuracy of the machine tool and made it possible to consistently produce accurate parts.

The control systems now in use are capable of ensuring that the machine tool will produce accurate parts every time. Factors such as cutter compensation are built into the MCU and make the necessary adjustments to compensate for cutting tool wear, ensuring that accuracy will be maintained.

Repeatability

Repeatability and reliability are very difficult to separate because many of the same variables affect each. Repeatability of a machine tool involves comparing each part produced on that machine to see how it compares to others for size and accuracy. The repeatability of an NC machine should be at least one-half the smallest tolerance allowed for on the part. Machine tools capable of greater accuracy and repeatability will naturally cost more because this accuracy must be built into the machine tool and/or the control system.

In order to maintain the repeatability of a machine tool, it is important that a regular maintenance program be established for each machine. Care in load-

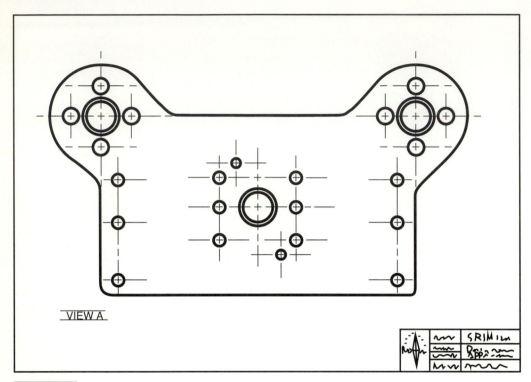

VIEW A

SRIM in

Fig. 4-3 A typical part which must be produced for a customer. (*Coleman Engineering Company*)

ing and clamping parts accurately in workholding devices such as fixtures and vises (whether manually or automatically by robots) is of prime importance, since an incorrectly held or positioned part will result in scrap work.

Productivity

The goal of industry has always been to produce better products at competitive or lower prices to gain a bigger share of the market. Soaring production worldwide has increased the competition for global and domestic markets. To meet this competition, manufacturers must continue to reduce manufacturing costs and build better-quality products, and they must get greater output per worker, greater output per machine, and greater output for each dollar of capital investment. These factors alone are justification for using NC and automating our factories. It provides us with the opportunity to produce goods of better quality, faster, and at a lower cost.

Let us compare the costs of producing a typical part manually and by NC. The part shown in Fig. 4-3 requires 25 holes of four different sizes, involving the operations of spotting, drilling, countersinking, counterboring, reaming, and drilling. Eleven different cutting tools are required for these operations, and the locational tolerance for the holes is ±0.001 in. (0.025 mm).

Let us assume that a firm received an order for 700 of these parts, which had to be supplied, as required by the customer, in lots of 35. Because the delivery date is uncertain, it is quite likely that the firm may produce these only as required, and would need 20 lots or manufacturing runs to complete the order. For a comparison of conventional operation with NC, see Fig. 4-4A and 4-4B and Table 4-1.

The savings of $8961 in this example is fairly typical and shows the increased productivity along with savings in production costs. Naturally it is difficult to make a general rule regarding the savings that NC offers; some jobs will show greater savings, while others will show less.

There are numerous other benefits of NC besides lower tooling costs and increased production. Some of the more common are:

1. *Reduced space for tool and fixture storage.* NC tapes take very little space and can be stored in a small cupboard or even a filing cabinet.

2. *Reduced parts inventory.* The NC tape can be reused as often as parts are required. This releases capital funds and reduces storage requirements.

3. *Lead time initially and for part design changes are reduced drastically.*

4. *Uniformity of finished parts.* There is very little loss for scrap parts because of the accuracy possible through NC.

Table 4-1	**COMPARISON OF JOB COSTS WITH AND WITHOUT NC**		
	Manually	**With NC**	**Savings**
Fixtures	Two fixtures required. Tool design, fabrication, inspection, and trial costs = $4200	One simple fixture required ($125) plus 2 hours (h) of tape preparation @ $20/h ($40) = $165	$4135
	Four- to six-week tooling lead time	One day for alignment of fixture and tape preparation	?
Floor-to-floor time/part	30 minutes (min) × 700 parts = 350 h @ $20/h = $7000	10 min × 700 parts = 117 h @ $20/h = $2340	$4660
Setup time/job lots	30 min × 20 lots = 10 h @ $20/h = $200	5 min × 20 lots = 1.7 h @ $20/h = $34	$166
		TOTAL SAVINGS	$8961+

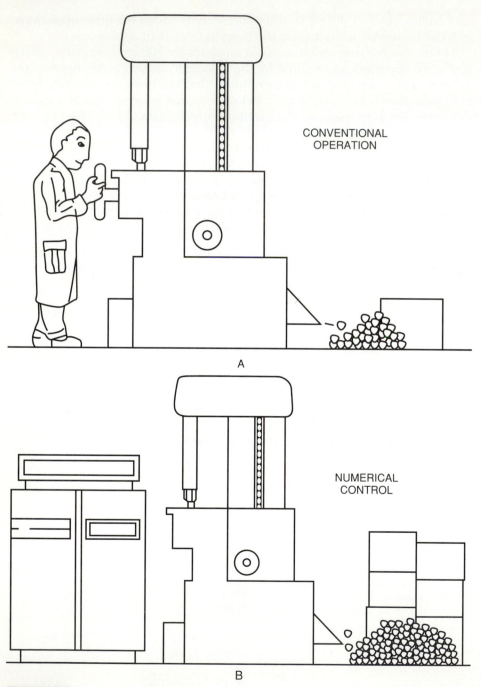

CONVENTIONAL
OPERATION

A

NUMERICAL
CONTROL

B

Fig. 4-4 Producing parts by means of (A) the conventional method versus (B) using NC, which results in increased production rates. (*Coleman Engineering Company*)

5. *Inspection time is reduced.* Tapes can be pretested and the machine accuracy checked periodically.

TAPE READERS

Machine tools themselves cannot read the coded commands contained on NC punched tape. The actual reading (decoding) is done by the tape reader mechanism, which is actually a part of the MCU (Fig. 4-5A and 4-5B). The purpose of the tape reader is to decode the information on the part program from the punched tape and send it to the MCU (Fig. 4-6).

Let us assume that the tape reader has read a dimension of X1.375. This is the path that this information follows in order for the machine tool to make the proper movement along the X axis:

1. An electrical signal created by holes or no holes in the punched tape comes from the tape reader to the *alphanumeric decoders*.

2. The *alpha* code, in this case the letter X, is read and decoded by the *alpha decoder* and is sent to the *register and buffer storage box*.

3. This box sends the information (X) from the buffer storage area to the X *stores* and leaves its top open to receive new information.

4. The *numeric* code, in this case the dimension 1.375, is read and decoded by the *numeric decoder* and sent directly to the X *stores*.

5. As the system operates, the signal for the dimension X1.375 is sent from the X *stores* to the positioning units of the motor, which will move the machine table 1.375 along the X axis.

FEEDBACK SYSTEMS

NC systems can read punched tape and direct a machine tool to perform a wide variety of operations. It is important that the machine tool perform these operations correctly and to the exact dimensions called for on the part print. In order to achieve this accuracy, there must be some method of checking the amount a machine table has moved against the amount it was asked to move by the MCU. *Feedback devices* are used to send data back to the MCU so that the machine table slide position can be compared with the required dimen-

A

Fig. 4-5A   A photoelectric tape reader. (*The Superior Electric Company*)

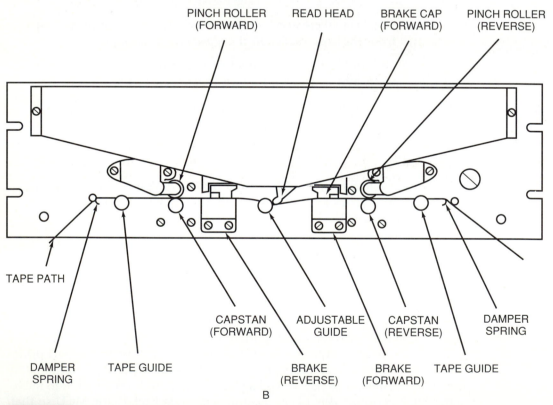

B

Fig. 4-5B   The photoelectric tape reader operates on the reflected light principle.

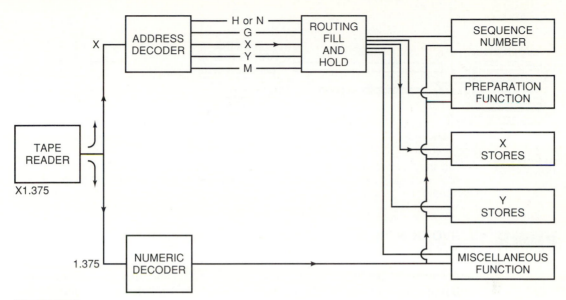

A block diagram of how the tape reader decodes the tape and sends the information to the proper storage areas.

sion of the input data. If any difference exists, the drive motor will be told (through electrical pulses) to make the necessary corrections.

There are two types of feedback systems—analog and digital. They use some form of feedback device on the machine slides or leadscrews to indicate the exact position of the machine table. Some of the more common feedback devices are Farrand scales or rotary pickup units.

Analog

Analog transducers, such as potentiometers and synchros, produce an electrical voltage which varies as the input shaft is turned or rotated (Fig. 4-7). This voltage is in proportion to the rotation of the input shaft which can be converted into very accurate machine table positions.

Digital

Digital feedback units, attached to the leadscrew of a machine tool, change the rotary motion of the machine screws to individual or discrete electrical pulses (Fig. 4-8). This series of pulses can be counted to indicate exactly how much the leadscrew shaft has turned, which indicates the amount the machine table has moved.

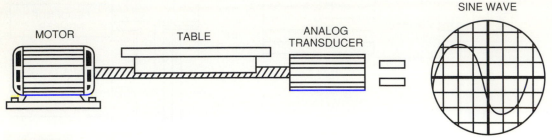

Fig. 4-7   Analog transducers produce an electrical voltage whose variations in level can be sensed, measured, and converted into accurate distances. (*Coleman Engineering Company*)

SERVO CONTROLS

Servo controls can be any group of electrical, hydraulic, or pneumatic devices which are used to control the position of machine tool slides. The most common servo control systems in use are the open-loop and the closed-loop systems.

Open-Loop System

In the *open-loop system* (Fig. 4-9), the tape is fed into a *tape reader* which decodes the information punched on the tape and stores it briefly until the machine is ready to use it. The tape reader then converts the information into electrical pulses or signals. These signals are sent to the *control unit*, which energizes the *servo control units*. The servo control units direct the *servomotors* to perform certain functions according to the information supplied by the tape. The amount each servomotor will move depends upon the number of

Fig. 4-8   Digital transducers produce electrical pulses which can be counted and converted into accurate distances. (*Coleman Engineering Company*)

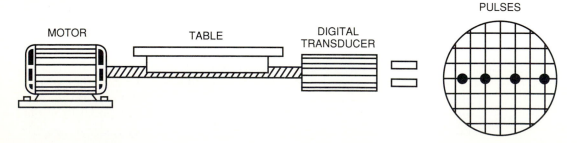

PULSES

MOTOR   TABLE   DIGITAL TRANSDUCER

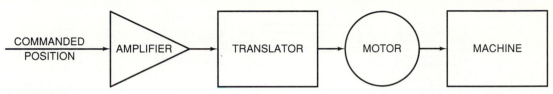

Fig. 4-9 Open-loop systems use unique stepping motors to move the machine table to the desired position.

electrical pulses it receives from the servo control unit. Precision leadscrews, usually having 10 threads per inch (tpi), are used on NC machines. If the servomotor connected to the leadscrew receives 1000 electrical pulses, the machine slide will move 1 in. (25.4 mm). Therefore, one pulse will cause the machine slide to move 0.001 in. (0.0254 mm). The open-loop system is fairly simple; however, since there is no means of checking whether the servomotor has performed its function correctly, it is not generally used where an accuracy greater than 0.001 in. (0.025 mm) is required.

The open-loop system may be compared to a gun crew that has made all the calculations necessary to hit a distant target but does not have an observer to confirm the accuracy of the shot.

Closed-Loop System

The *closed-loop system* (Fig. 4-10) can be compared to the same gun crew that now has an observer to confirm the accuracy of the shot. The observer relays the information regarding the accuracy of the shot to the gun crew, which then makes the necessary adjustments to hit the target.

The closed-loop system is similar to the open-loop system with the exception that a *feedback unit* (Fig. 4-10) is introduced into the electrical circuit. This feedback unit, often called a *transducer*, compares the amount the machine table has been moved by the servomotor with the signal sent by the control unit. The control unit instructs the servomotor to make whatever adjustments are necessary until both the signal from the control unit and the one from the servo unit are equal. In the closed-loop system, 10,000 electrical pulses are required to move the machine slide 1 in. (25 mm). Therefore, on this type of system, one pulse will cause a 0.0001-in. (0.0025-mm) movement of the machine slide. Closed-loop NC systems are very accurate because the command signal is recorded, and there is an automatic compensation for error. If the machine slide is forced out of position due to cutting forces, the feedback unit indicates this movement and the machine control unit (MCU) automatically makes the necessary adjustments to bring the machine slide back to position.

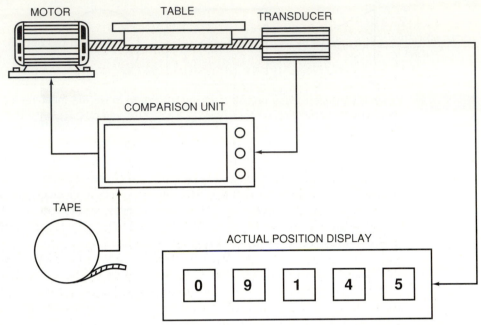

MOTOR   TABLE   TRANSDUCER

COMPARISON UNIT

TAPE

ACTUAL POSITION DISPLAY

| 0 | 9 | 1 | 4 | 5 |

Fig. 4-10 Closed-loop systems contain some type of feedback device to make sure that the machine table is in the exact position called for by the MCU. (*Coleman Engineering Company*)

MCU

The MCU (Fig. 4-11) is the intermediary in the total NC operation. Its main function is to take the part program and convert this information into a language that the machine tool can understand so that it can perform the functions required to produce a finished part. This could include turning relays or solenoids ON or OFF and controlling the machine tool movements through electrical or hydraulic servomechanisms.

Data Decoding and Control

One of the first operations that the MCU must perform is to take the binary-coded data (BCD) from the punched tape and change it into binary digits. This information is then sent to a holding area, generally called the *buffer storage*, of the MCU (Fig. 4-12). The purpose of the buffer area is to allow the information or data to be transferred faster to other areas of the MCU. If there were no buffer storage, the MCU would have to wait until the tape reader decoded and sent the next set of instructions. This would cause slight pauses in the transfer of information, which in turn would result in a pause in the

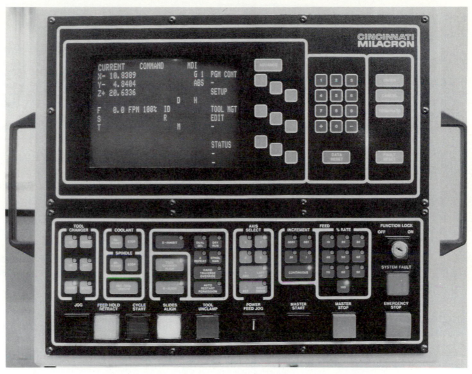

Fig. 4-11   The MCU is the intermediary in the total NC operation. (*Cincinnati Milacron, Inc.*)

machine tool motion and cause tool marks in the workpiece. MCUs which do not have buffer storage must have high-speed tape readers to avoid the pauses in transferring information and the machining operation.

The data decoding and control area of the MCU processes information which controls all machine tool motions as directed by the punched tape. This area also allows the operator to stop or make changes to the program manually, through the control panel.

MCU Development

Since the early 1950s, MCUs have developed from the bulky vacuum tube units to today's computer control units, which incorporate the latest microprocessor technology. Until the early 1970s, all MCU functions—such as tape format recognition, absolute and incremental positioning, interpolation, and code recognition—were determined by the electronic elements of the MCU. This type of MCU was called *hard-wired* because the functions were built into the computer elements of the MCU and could not be changed.

The development of *soft-wired* controls in the mid-1970s resulted in more

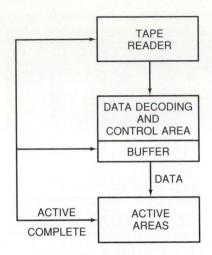
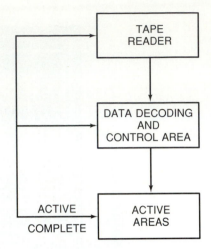

When active areas complete current block, new data is transferred from buffer, and the tape reader starts to refill the buffer with new tape data

When active areas complete current block, the tape reader starts to read into the active areas the new data

Fig. 4-12   The buffer storage transfers information quickly to other areas of the MCU to ensure continuous machine tool motion.

flexible and less costly MCUs. Simple types of computer elements, and even minicomputers, became part of the MCU. The functions that were locked in by the manufacturer with the early hard-wired systems are now included in the computer software within the MCU. This *computer logic* has more capabilities, is less expensive, and at the same time can be programmed for a variety of functions whenever required.

CPU

The CPU of any computer contains three sections: control, arithmetic-logic, and memory. The CPU control section is the workhorse of the computer (Fig. 4-13). Some of the main functions of each part of the CPU are as follows:

1. *Control section*
   - Coordinates and manages the whole computer system.
   - Obtains and decodes data from the program held in memory.
   - Sends signals to other units of the NC system to perform certain operations.

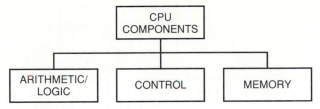

Fig. 4-13 The CPU is the workhorse of any computer.

2. *Arithmetic-Logic*
- Makes all calculations, such as adding, subtracting, multiplying, and counting as required by the program.
- Provides answers to logic problems (comparisons, decisions, etc.).

3. *Memory*
- Provides short-term or temporary storage of data being processed.
- Speeds the transfer of information from the main memory of the computer.
- Has a memory register which provides a specific location to store a word and/or recall a word.

COMPUTER NUMERICAL CONTROL

In 1970, a new term—computerized numerical control (CNC)—was introduced to the NC vocabulary. The development of an CNC system was made economically possible by electronic breakthroughs which resulted in lower costs for minicomputers.

The physical components of CNC or *soft-wired* units are the same regardless of the type of machine tool being controlled. On soft-wired units, it is not the MCU but the *executive program*, or *load tape*, which is loaded into the CNC computer's memory by the control manufacturer. It is this executive program that makes the control unit "think" like a turning center or a machining center. Therefore, if there is a need to change the functions of a soft-wired or CNC unit, the executive program can be changed to suit. Usually it is the manufacturer, and not the user, who makes the change.

There have been great developments in MCUs with the introduction of large-scale ICs and microprocessors. Many features not available a few years ago give much greater flexibility and productivity to NC systems. As developments in control units continue, new features will become available which will make the machines they control more productive.

The CNC machine control unit of today has several features which were not found on the pre-1970 hard-wired control units (Fig. 4-14).

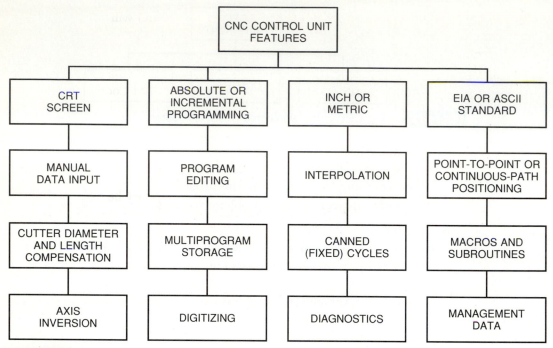

Fig. 4-14 The CNC units of today contain many features not found on earlier control units.

1. *Cathode-Ray Tube (CRT)*
   The CRT is like a TV screen which serves a number of purposes:
   a. It shows the exact position of the machine table and/or the cutting tool at every position while a part is being machined.
   b. The entire program for a part can be displayed on the screen for editing or revision purposes.
   c. The screen assists in work setups, and some models receive messages from sensors which indicate machine or control problems.

2. *Absolute and Incremental Programming*
   By using the proper address code G90 (absolute) or G91 (incremental), most CNC units will automatically program in that particular mode. Some of the newer CNC units are capable of handling mixed data (absolute and incremental) in a given block of data.

3. *Inch or Metric*
   Most CNC units are capable of working in inch or metric dimensions. Either a switch on the control unit or a specific code on the

part program (G70 for inch and G71 for metric) will determine the measurement system used when machining a part.

4. *EIA or ASCII Code*
   Many of the newer CNC units will read either the Electronic Industries Association (EIA) or the American Standard Code for Information Interchange (ASCII) standard code. The control unit identifies each one by the parity check: odd for the EIA code and even for the ASCII code.

5. *Manual Data Input*
   Most CNC units provide some method of making changes to the part program, if necessary. This may be necessary because changes were made to the part specifications, to correct an error, or to change the machining sequences of the part.

6. *Program Editing*
   Very few part programs are free from error from the start, and the flaws generally show up on the shop floor. Program editing is a feature which allows the part program to be corrected or changed right at the control unit.

7. *Interpolation*
   While early models of control units were capable of only linear, circular, or parabolic interpolation, the newer models include helical and cubic interpolation.

8. *Point-to-Point and Continuous-Path Positioning*
   All MCUs are capable of point-to-point or continuous-path positioning or any combination of each.

9. *Cutter Diameter and Length Compensation*
   On newer MCUs, it is possible to manually enter the diameter and/or the length of a cutter which may vary from the specification of the part program. The control unit automatically calculates what adjustments are necessary for the differences in size and moves the necessary slides to adjust for the difference.

10. *Program Storage*
    The newer CNC units generally provide large-capacity computing and data storage (memory). This allows information about a part program to be entered (manually or from tape) and stored for future use. Therefore, whenever the part program is required, it can be recalled from memory rather than reread from the tape. This not only protects the quality of the tape, but the information is recalled much faster.

11. *Canned or Fixed Cycles*

Storage capacity is generally provided in the MCU for any cycle (machining, positioning, etc.) which is used or repeated in a program. When the program is being written and a previous cycle must be repeated, all the programmer has to do is insert a code in the program where that cycle is required. The control unit will recognize that code and recall from memory all the information required to perform that cycle (operation) again.

12. *Subroutines and Macros*

A parametric subroutine, sometimes called a "program within a program," is used to store frequently used data sequences (one block or a number of blocks of information) which can be recalled from memory as required by a code in the main part program (Fig. 4-15). An example of a subroutine could be a drilling cycle in which

Fig. 4-15 Subroutines consist of a block or blocks of information which are stored in memory and recalled whenever they are required. (*The Superior Electric Company*)

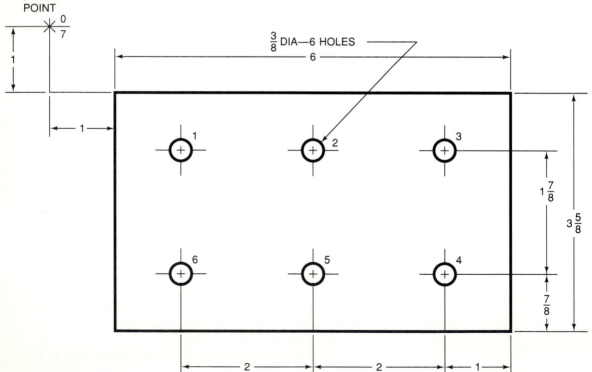

a series of ⅜-in.-diameter (9.5-mm) holes 1 in. (25 mm) deep must be drilled in a number of locations on a workpiece.

A *macro* is a group of instructions or data which are permanently stored and can be recalled as a group to *solve recurring problems* such as bolt-hole circle locations, drilling and tapping cycles, and other frequently used routines. An example of a macro would be the XY locations of various holes on any bolt-circle diameter. When the diameter of the bolt circle and the number of holes on the circle are provided, the MCU makes all the calculations for hole locations and causes the machine tool slides to move into the proper position for each hole.

13. *Axis Inversion (Mirror Image)*
Axis inversion (Fig. 4-16) is the ability of the MCU to reverse plus and minus (+ and −) values along an axis or program zero to produce an accurate left-hand part from a right-hand program. This ability (called *symmetrical machining*) applies to all four quandrants and greatly reduces the time that would be required to program each part.

14. *Digitizing*
The digitizing feature allows a part program to be made directly from an existing part. The original part is traced while the CNC unit records all machine motions and produces a part program on punched tape.

15. *Diagnostics*
Diagnostic capabilities (on tape or built-in) can monitor all conditions and functions of an NC machine or the control unit. If an error

Fig. 4-16 Axis inversion (mirror image) can be used to produce either left- or right-hand parts from one program. (*The Superior Electric Company*)

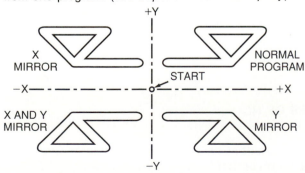

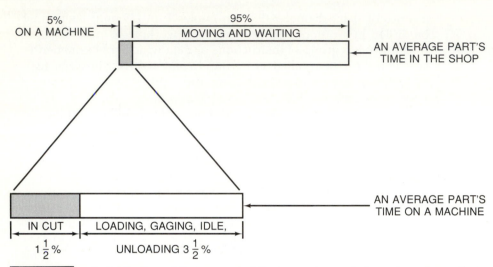

Fig. 4-17 A distribution of the amount of time a part spends in a shop shows that only a small percentage of time is spent actually machining the part. (*Cincinnati Milacron, Inc.*)

or malfunction occurs or if a changing condition nears a critical point, a signal or message is shown on the CRT. On some control units, as a critical point gets close, there may be a warning to the operator, shown on the screen, or the machine may automatically shut down.

Diagnostic software routines can be used to check hardware modules, circuit boards, and every area of the control unit to check their accuracy. The information is then displayed on the CRT.

16. *Management Data*

The modern MCU controls almost all machine tool functions through the built-in computer. Since this information or data is already in the MCU controller or computer, it can be sent to the host or mainframe computer to provide valuable data to management and the operator. Spindle-on time, part-run time, number of parts machined, etc., can be recorded and sent to the host computer or displayed on the CRT screen.

ADVANTAGES OF NC

Recent studies show that of the amount of time an average part spent in a shop, only a fraction of that time was actually spent in the machining process. Figure 4-17 shows the distribution of time in a shop where parts are machined in small batches. Let us assume that a part spent 50 hours from the time it arrived at a plant as a rough casting or bar stock to the time it was a finished product. During this time, it would be on the machine for only 2½ hours and be cut for only ¾ hour. The rest of the time would be spent on waiting, moving, setting up, loading, unloading, inspecting the part, setting speeds and feeds, and changing cutting tools.

NC reduces the amount of non-chip-producing time by selecting speeds and feeds, making rapid moves between surfaces to be cut, using automatic fixtures, automatic tool changing, controlling the coolant, in-process gaging, and loading and unloading the part. These factors, plus the fact that it is no longer necessary to train machine operators, have resulted in considerable savings throughout the entire manufacturing process and caused tremendous growth in the use of NC. Some of the major advantages of NC are as follows:

1. There is automatic or semiautomatic operation of machine tools. The degree of automation can be selected as required.

2. Flexible manufacturing of parts is much easier. Only the tape needs changing to produce another part.

3. Storage space is reduced. Simple workholding fixtures are generally used, reducing the number of jigs or fixtures which must be stored.

4. Small part lots can be run economically. Often a single part can be produced more quickly and better by NC.

5. Nonproductive time is reduced. More time is spent on machining the part, and less time is spent on moving and waiting.

6. Tooling costs are reduced. In most cases complex jigs and fixtures are not required.

7. Inspection and assembly costs are lower. The quality of the product is improved, reducing the need for inspection and ensuring that parts fit as required.

8. Machine utilization time is increased. There is less time that a machine tool is idle because workpiece and tool changes are rapid and automatic.

9. Complex forms can easily be machined. The new control unit features and programming capabilities make the machining of contours and complex forms very easy.

10. Parts inventory is reduced. Parts can be made as required from the information on the punched tape.

Since the first industrial revolution, about 200 years ago, NC has had a significant effect on the industrial world. The developments in the computer and NC have extended a person's mind and muscle. The NC unit takes symbolic input and changes it to useful output, expanding a person's concepts into creative and productive results. NC technology has made such rapid advances that it is being used in almost every area of manufacturing, such as machining, welding, press-working, and assembly.

If industry's planning and logic are good, the second industrial revolution will have as much or more effect on society as the first industrial revolution had. As we progress through the various stages of NC, it is the entire manufacturing process which must be kept in mind. Computer-assisted manufacturing (CAM) and computer-integrated machining (CIM) are certainly where the future of manufacturing lies, and considering the developments of the past, it will not be too far in the future before the automated factory is a reality.

REVIEW QUESTIONS

1. Name five key elements which are involved from the editing of punched tape to producing a finished part.

NC Performance

2. List four important developments which had an important effect on the great advances made by NC.

3. To what accuracy are NC machine tools capable of working? Explain why.

4. What improvements have been made to machine tools to make them more reliable?

5. How were control systems improved to ensure that accurate products were produced?

6. Define the term *repeatability*.

7. Name three factors which alone are justification enough for using NC.

8. Compare the manual machining of a part with NC machining with regard to fixtures, floor-to-floor time, setup time, and dollar savings.

Tape Readers

9. What is the purpose of a tape reader?

10. Explain what occurs when a tape reader reads a dimension of Y2.125.

Feedback Systems

11. What is the purpose of feedback devices?

12. Name two types of feedback systems and state how each operates.

Servo Controls

13. Briefly explain how an open-loop system operates.

14. Compare the closed-loop system with the open-loop system and state its advantages.

15. How many electrical pulses are required to move a machine slide 1 in. (25 mm) on a closed-loop system?

MCU

16. What is the purpose of the MCU?

17. Explain what purpose the buffer storage of the MCU serves.

18. What is the purpose of the data decoding and control section of the MCU?

19. Explain the difference between hard-wired and soft-wired controls.

20. What purposes does the CPU serve?

21. List two important functions of each of the CPU sections.

CNC

22. Explain the purpose of the executive program.

23. What is the purpose of the CRT?

24. How can the MCU recognize absolute or incremental programming?

25. What is the purpose of cutter diameter and length compensation?

26. Briefly explain:
  (a) Canned or fixed cycles
  (b) Subroutines and macros
  (c) Digitizing

27. What is the purpose of the axis inversion feature?

Advantages of NC

28. If an average part spends about 50 hours in a shop, how much time is likely spent in actually machining the part? How was the rest of the time most likely spent?

29. Why has NC been so widely accepted by industry?

30. List five of the most important advantages of NC.

CHAPTER

FIVE

Programming Data

The programmer's function is to take information from an engineering drawing and convert it into data which the control unit of the machine tool will understand. In order to do this successfully, a programmer must have a good knowledge of machining operations and sequences, workholding methods, metal properties, cutting tools, and cutting speeds and feeds. The programmer must also understand the various codes and functions used in numerical control (NC) so that accurate parts can be produced in the minimum amount of time. It is very important that there be a close liaison between the product engineer, the drafter, and the machine tool operator to ensure that the correct product is produced to the proper size.

After completing
this chapter,
you should be
able to:

1. Understand the various codes used in NC

2. Understand the purpose of the various NC functions

3. Identify the correct speeds and feeds to be used for cutting various metals

NC FUNCTIONS

NC functions have been standardized by the Electronic Industries Association (EIA) and the American Standard Code for Information Interchange (ASCII). This coding system is used by most NC machine and control manufacturers. Although there are slight variations between each system, the function codes are basically the same.

Various types of function codes are used in NC work to indicate the type of operation that the machine tool is to perform. The most common are the *preparatory,* or G, functions, the *miscellaneous,* or M, functions, and the *address character,* or A to W, functions which must appear on the manuscript and the NC tape.

NC programming is done in a variable-block format with word (letter) addressing. Each instruction word consists of an address character (X, Y, Z, F, or M) followed by the numerical data. The letter address (code) tells the machine control unit (MCU) what machine function to perform, while the number code usually gives the distance, feed rate, speed, etc.

Let us examine a sample block of NC information in order to understand programming sequence (Fig. 5-1).

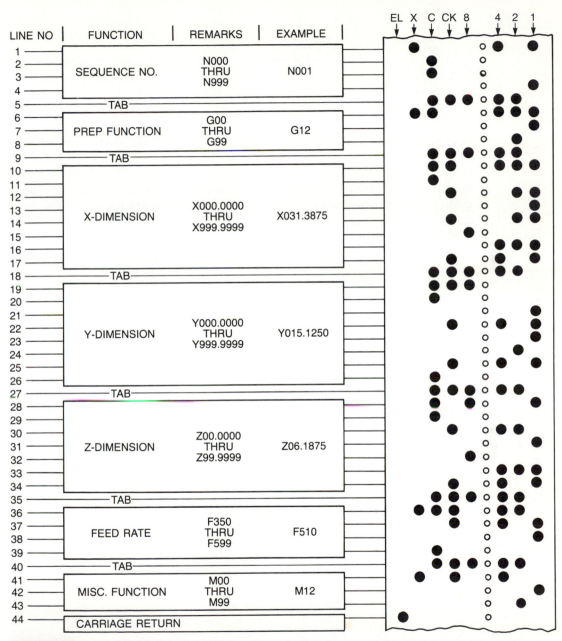

Fig. 5-1   EIA interchangeable variable block format (word address tab ignore). (*Cincinnati Milacron, Inc.*)

SEQUENCE NUMBER

The first word in a block of information is usually the *sequence number,* and it is used to identify each particular block of information. The sequence number is usually a three-digit numerical word (though it can also be four digits) preceded by the letter code N, H, or O. It is common practice, but not essential, to use an H or O for the first block of information in every program, after each cutting tool change, and for aligning and realigning blocks. The letter N is generally used to identify all other sequence blocks. Sequence numbers range from N000 to N9999. Sequence numbers are especially valuable when it becomes necessary to edit a tape and make corrections or revisions to a program. They can be used to search the tape or program for a particular sequence number, stop when it is found, and display it on the cathode-ray tube (CRT) screen for editing purposes.

If sequence numbers are assigned with flexibility in mind during programming, they can be very useful if it becomes necessary to add, revise, or correct a program. It is not considered good practice to assign numbers in numerical order (for example, N001, N002, N003), because this approach provides no flexibility if revisions or additions must be made to a program. If a correction or revision must be made to the program, all following sequence numbers must be renumbered. The most common method used to assign sequence numbers is in progression of 10 (for example, N010, N020, N030). This allows enough room between each sequence number to insert as many as nine pieces of new information before any sequence numbers would have to be renumbered. The following illustrates the numerical order approach.

N001

New sequence inserted must become N002.

N002

All following sequences must be renumbered.

The following illustrates the progression of 10 system.

N010

New sequences N011 to N019 can be inserted.

N020

No renumbering required.

N100

PREPARATORY FUNCTIONS

The *preparatory function* or *cycle code* refers to some mode of operation of the machine tool or NC system. It generally refers to some action on the X, Y, and/or Z axes. It is generally common practice for the X and Y axes to be positioned before the Z axis is activated. In NC programming, the word address letter G refers to a preparatory function and is followed by a two-digit number; e.g., a G00 function would be point-to-point positioning at a rapid rate of about 150 to 400 in./min (38 to 100 m/min). Some common preparatory functions include such operations as point-to-point positioning, linear interpolation, parabolic interpolation, absolute or incremental programming, inch or metric programming, and fixed (canned) cycles. Each of these is designated by a G code, which the central processing unit (CPU) and the machine tool can recognize and act upon accordingly.

    Fixed or *canned cycles*, identified by the preparatory function codes G81 to G89, are a *preset combination of operations* which cause the machine axis movement and/or cause the machine spindle to complete such operations as drilling, boring, and tapping. Each operation may involve as many as seven machine movements, and a G81 to G89 code is all that is required on the tape program to make the machine perform a particular operation. Control units with this feature can save up to 50 percent in programming time and one-third of the data processing time, and reduce the length of tape required for the program.

    Most manufacturers of NC machines and control systems use the standard EIA numbering systems for fixed or canned cycles. An NC turret-drilling machine will be used to illustrate and describe some of the more common fixed cycles.

Drill Cycle (G81) (Fig. 5-2)

In sequence number (H060), a drill cycle (G81) is used to drill a 1-in.-deep hole (Z1.225) at 11 in./min (F511) at position #1.

1. First the spindle "rapids" to gage height at level #1; then it feeds to depth at level #2 (Z1.225).

2. The spindle will rapid back to gage height (level #1).

3. In sequence N070, the table will move (X08000) along the X axis to position #2 and repeat the cycle.

To calculate the programmed Z depth (level #2) for a 118° included angle drill

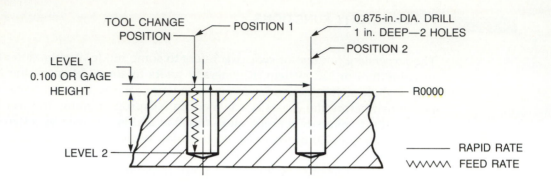

SEQ. NO.	PREP FUNCT.	X POSITION		Y POSITION		Z FEED POSITION		Z FEED	Z RAPID POSITION		SPINDLE SPEED	TOOL FUNCTION	MISC FUNCT	POS NO
H061	G81	X06	000	Y04	000	Z1	263	F511	R0	000	S635	T4	M13	1
N062		X08	000										M06	2

Fig. 5-2 A drill cycle to drill two 0.750-in.-diameter holes 1 in. deep. (*Cincinnati Milacron, Inc.*)

point, use the following formula:

$$Z = \text{Full body depth} + \text{drill point length}$$
$$= 1.000 + (0.300 \times \text{drill diameter})$$
$$= 1.000 + (0.300 \times 0.750)$$
$$= 1.000 + 0.225$$
$$= 1.225$$

Drill/Dwell Cycle (G82) (Fig. 5-3)

In sequence H080, code G82 calls for a dwell cycle. This cycle is the same as a drill cycle with a dwell time added at the programmed depth.

1. The spindle feeds from the gage height (level #1) to the depth at level #2.

2. It stops at the depth for a period of time which is preset on the dwell timer, and then it finishes the cycle.

3. In sequence N100, the table moves along the X axis (X08000) to position #2.

4. The cycle is repeated at position #2.

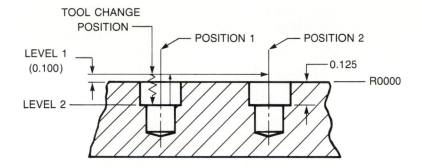

SEQ. NO.	PREP. FUNCT.	X POSITION	Y POSITION	Z FEED POSITION	Z FEED	Z RAPID POSITION	SPINDLE SPEED	TOOL FUNCTION	MISC FUNCT	POS. NO.
H063	G82	X06 ¦ 000	Y04 ¦ 000	Z0 ¦ 125	F450	R0 ¦ 000	S615	T5	M13	1
N064		X08 ¦ 000								2

Fig. 5-3   A dwell cycle produces a good finish on holes requiring spot facing or counterboring. (*Cincinnati Milacron, Inc.*)

Note: The dwell cycle is generally used for operations such as counterboring and spot facing where a smooth surface is required.

Tap Cycle (G84) (Fig. 5-4)

In sequence H120, code G84 calls for a tap cycle to tap a hole 1.000 in. deep (Z1.000) at 31 in./min (F531).

1. The spindle feeds from gage height at level #1 to the depth at level #2.

2. At 1.000 in. depth, the spindle reverses and feeds back up to gage height (level #1).

3. The spindle then reverses direction again.

4. In sequence N130, the table moves (09000) along the Y axis to position #2, and the next hole is tapped.

Bore Cycle (G85) (Fig. 5-5)

In sequence #H040, code G85 calls for a bore cycle to bore a 1.250-in. hole at 2 in./min.

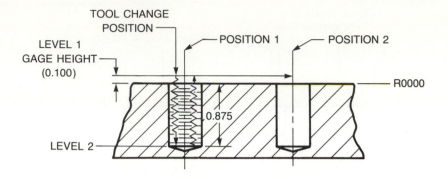

SEQ. NO.	PREP. FUNCT.	X POSITION		Y POSITION		Z FEED POSITION		Z FEED	Z RAPID POSITION		SPINDLE SPEED	TOOL FUNCTION	MISC. FUNCT.	POS. NO.
H080	G84	X15	000	Y10	000	Z0	875	F531	R0	000	S635	T2	M03	1
N081				Y09	000								M06	2

Fig. 5-4   A tap cycle used to thread two holes 1 in. deep. (*Cincinnati Milacron, Inc.*)

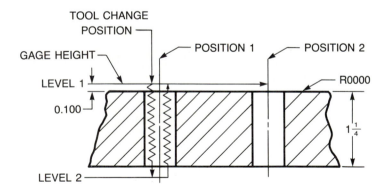

SEQ. NO.	PREP. FUNCT.	X POSITION		Y POSITION		Z FEED POSITION		Z FEED	Z RAPID POSITION		SPINDLE SPEED	TOOL FUNCTION	MISC. FUNCT.	POS. NO.
H011	G85	X28	000	Y18	000	Z1	250	F420	R0	000	S675	T6	M03	1
N012		X24	000	Y14	000									2

Fig. 5-5   The bore cycle is used to bring a hole to accurate size and location. (*Cincinnati Milacron, Inc.*)

1. The table positions in the X and Y axes, and then the spindle rapids to level #1.

2. The boring tool is then fed to depth at level #2.

3. The direction of feed reverses, and the boring tool returns to level #2.

4. In sequence N050, the table moves along the X and Y axes to position #2, and the cycle is repeated.

The following preparatory functions have been supplied by the EIA and are according to their standard E1A-274-D.

Number	Operation	Definition
G00	Point-to-Point Positioning	Point-to-point positioning at rapid or other traverse rate.
G01	Linear Interpolation	A mode of contouring control which uses the information contained in a block to produce a straight line in which the vectorial velocity is held constant.
G02	Arc Clockwise (Two-Dimensional)	An arc generated by the coordinated motion of two axes in which curvature of the path of the tool with respect to the workpiece is clockwise, when viewing the plane of motion in the negative direction of the perpendicular axis.
G03	Arc Counterclockwise (Two-Dimensional)	An arc generated by the coordinated motion of two axes in which curvature of the path of the tool with respect to the workpiece is counterclockwise, when viewing the plane of motion in the negative direction of the perpendicular axis.
G02–G03	Circular Interpolation (Two-Dimensional)	A mode of contouring control which uses the information contained in a single block to produce an arc of a circle. The

Number	Operation	Definition
		velocities of the axes used to generate this arc are varied by the control.
G04	Dwell	A timed delay of programmed or established duration, not cyclic or sequential; i.e., not an interlock or hold.
G06	Parabolic Interpolation	A mode of interpolation used in contouring to produce a segment of a parabola. Velocities of the axes used to generate this curve are varied by the control.
G08	Acceleration	A controlled velocity increase to programmed rate starting immediately.
G09	Deceleration	A controlled velocity decrease to a fixed percent of the programmed rate starting immediately.
G13–G16	Axis Selection	Used to direct a control to the axis or axes, as specified by the Format Classification, as in a system which time-shared the controls.
G17–G19	Plane Selection	Used to identify the plane for such functions as circular interpolation, cutter compensation, and others as required.
G33	Thread Cutting, Constant Lead	Mode selection for machines equipped for thread cutting.
G34	Thread Cutting, Increasing Lead	Mode selection for machines equipped for thread cutting where a constantly increasing lead is desired.
G35	Thread Cutting, Decreasing Lead	Mode selection for machines equipped for thread cutting where a constantly decreasing lead is desired.

Number	Operation	Definition
G40	Cutter Compensation/Offset Cancel	Command which will discontinue any cutter compensation/offset.
G41	Cutter Compensation—Left	Cutter on left side of work surface looking from cutter in the direction of relative cutter motion with displacement normal to the cutter path to adjust for the difference between actual and programmed cutter radii or diameters.
G42	Cutter Compensation—Right	Cutter on right side of work surface looking from cutter in the direction of relative cutter motion with displacement normal to the cutter path to adjust for the difference between actual and programmed cutter radii or diameters.
G43	Cutter Offset— Inside Corner	Displacement normal to cutter path to adjust for the difference between actual and programmed cutter radii or diameters. Cutter on inside corner.
G44	Cutter Offset— Outside Corner	Displacement normal to cutter path to adjust for the difference between actual and programmed cutter radii or diameters. Cutter on outside corner.
G50–G59	Adaptive Control	Reserved for adaptive control requirements.
G70	Inch Programming	Mode for programming in inch units. It is recommended that control turn on establish this mode of operation.
G71	Metric Programming	Mode for programming in metric units. This mode is canceled by G70, M02, and M30.

Number	Operation	Definition
G72	Arc Clockwise (Three-Dimensional)	An arc generated by the coordinated motion of three axes in which the curvature of the tool path with respect to the workpiece is clockwise.
G73	Arc Counterclockwise (Three-Dimensional)	An arc generated by the coordinated motion of three axes in which the curvature of the tool path with respect to the workpiece is counterclockwise.
G72–G73	Circular Interpolation (Three-Dimensional)	A mode of contouring control which uses the information contained in a single block to produce an arc on a sphere. The velocities of the axes used to generate this arc are varied by the control.
G75	Multiquadrant Circular	MODE Selection if required for Multiquadrant Circular, cancelled by G74.
G80		Command that will discontinue any of the fixed cycles G81–G89.
G81–G89	Fixed Cycle	A preset series of operations which direct machine axis movement and/or cause spindle operation to complete such action as boring, drilling, tapping, or combinations thereof.
G90	Absolute Input	A control mode in which the data input is in the form of absolute dimensions.
G91	Incremental Input	A control mode in which the data input is in the form of incremental data.
G92	Preload of Registers	Used to preload registers to desired values. No machine operation is initiated. Examples would include preload of axis position registers, spindle speed

Fixed cycle			At bottom		Movement out	
Number	**Code**	**Movement in**	**Dwell**	**Spindle**	**to feed start**	**Typical usage**
1	G81	Feed	—	—	Rapid	Drill, Spot Drill
2	G82	Feed	Yes	—	Rapid	Drill, Counterbore
3	G83	Intermittent	—	—	Rapid	Deep Hole
4	G84	Spindle Forward Feed	—	Rev.	Feed	Tap
5	G85	Feed	—	—	Feed	Bore
6	G86	Start Spindle, Feed	—	Stop	Rapid	Bore
7	G87	Start Spindle, Feed	—	Stop	Manual	Bore
8	G88	Start Spindle, Feed	Yes	Stop	Manual	Bore
9	G89	Feed	Yes	—	Feed	Bore

Number	*Operation*	*Definition*
		constraints, initial radius, etc. Information within this block shall conform to the character assignments of the preceding table.
G93	Inverse Time Feedrate	The data following the feedrate address is equal to the reciprocal of the time in minutes to execute the blocks and is equivalent to the velocity of any axis divided by the corresponding programmed increment.
G94	Inches (Millimeters) Per Minute Feedrate	The feedrate code units are inches per minute or millimeters per minute.
G95	Inches (Millimeters) Per Revolution	The feedrate code units are inches (millimeters) per revolution of the spindle.
G96	Constant Surface Speed Per Minute	The spindle speed code units are surface feet (meters) per minute and specify the tangential surface speed of the tool relative to the

Number	Operation	Definition
		workpiece. The spindle speed is automatically controlled to maintain the programmed value.
G97	Revolutions Per Minute	The spindle speed is defined by the spindle speed word.

MACHINE MOTION AXES

All machine tools have sliding linear and rotating motion. The single-spindle NC drilling machine (Fig. 5-6) has three linear motions (two horizontal and one vertical) and one rotary motion (the spindle). NC makes it very important to identify these motions accurately so that they can be programmed and dimensioned as required. The rectangular coordinates (X, Y, and Z) make it

Fig. 5-6   The X, Y, and Z axes of an NC drilling machine give three-dimensional coordinates, with the Z axis in a vertical relationship to the XY axes. (*Cincinnati Milacron, Inc.*)

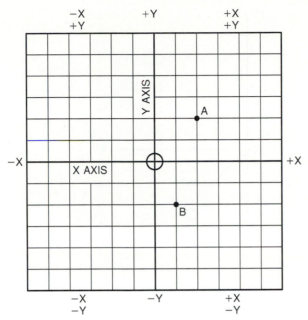

Fig. 5-7 Locating points within the XY coordinates. (*Allen-Bradley*)

possible to state the exact location of any point on a single plane or in a three-dimensional space.

The location of any point on a flat surface can be exactly positioned by making reference to where it is in relation to the X and Y axes (Fig. 5-7). The rectangular coordinates also allow any point in space to be located in relation to three perpendicular planes (XYZ) axes.

X, Y, and Z Words

X, Y, and Z words refer to coordinate movement of the machine tool for positioning or machining purposes. When points are located on a workpiece, two intersecting lines (one horizontal and the other vertical) at right angles to each other are used. These lines are called "axes," and where they intersect is called the "origin," or "zero point" (Fig. 5-8). The horizontal line is called the X axis, and the vertical line is called the Y axis.

To avoid confusion and errors when programming, the distance to the right or left of the Y axis is generally given first. Any X distance to the right of the Y axis is uniformly referred to as a positive (+) dimension; those to the left are negative (−) dimensions. Any Y distance above the X axis is referred to as a positive (+) dimension; those below are negative (−) dimensions. The third

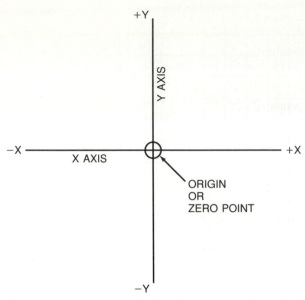

+Y

Y AXIS

−X ——————— X AXIS ——————— +X

ORIGIN
OR
ZERO POINT

−Y

Fig. 5-8   Intersecting horizontal and vertical lines form a right angle. (*Allen-Bradley*)

plane, or Z axis, is perpendicular to the plane established by the X and Y axes and on a vertical mill or drill press refers to the movement of the cutting tool. Toward or into the workpiece is a negative (−) motion, while away from the workpiece is a positive (+) motion.

Coordinate information must be programmed in the following sequence in order for the MCU to understand the command correctly.

1. *Axis Movement*
   X, Y, or Z axes must be specified. When motion is required on more than one axis, program X first, Y second, and Z last.

2. *Direction Movement*
   The information must indicate whether the movement is positive (+) or negative (−) from the origin point. The positive movement sign (+) does not have to appear on a program, since it is assumed. Negative dimensions must have the minus (−) sign. For example, the dimension X0068750 or X6.875 would mean a 6- and 875-thousandths-of-an-inch movement to the right of the Y axis.

3. *Dimension Movement*
   This is normally a seven-digit number with the decimal point fixed in

the tape format to allow four places to the right of the decimal. Decimal points are not usually programmed; however, modern controls do allow decimal points to be programmed. For example, the dimension Y − 0068750 or Y − 6.875 would mean a 6- and 875-thousandths-of-an-inch movement below the X axis.

4. *Depth Selection (Z Axis)*
Machining operations such as producing holes, milling slots or steps, or cutting into the surface of a workpiece generally involve the Z axis. The Z axis on any machine tool is usually a line drawn through the center of the machine spindle. Figure 5-6 shows the Z axis in relation to the X and Y axes of the machine table and the workpiece. The Z axis of motion is always parallel to the spindle of the machine and perpendicular to the workholding surface. On milling, boring, drilling and tapping machines, the spindle is the tool-rotating device. On lathes, grinders, and other machines where the work revolves, the spindle is the work-rotating device. A positive (+) Z movement moves the cutting tool *away* from the workpiece, while a negative (−) Z movement moves the cutting tool *into* a workpiece. The Z motion can be controlled by the operator (manual data input), by preset stops, or by programmed NC tape.

The word address letter Z generally consists of a seven-digit number with the decimal point fixed at four places to the left. The Z movement generally involves a rapid move to a programmed gage height or a position above the workpiece and then a slower feed rate during the machining operation. The following formula can be used if the entire Z axis movement must be programmed (Fig. 5-9).

$$Z = PS + CL + TL$$

Z = the distance from Z0 (zero) to the spindle gage line
PS = the distance from Z0 to the part surface
CL = the clearance
TL = the tool set length (from the spindle gage line to the cutting edge)

Work Plane

The word address letter R refers to either the work surface or the rapid-feed distance (sometimes called the *work plane*) programmed. The R work surface is set either at the highest surface on the part to be machined or at a specific

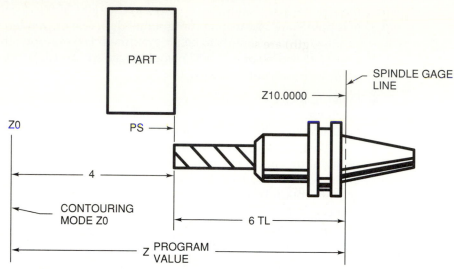

Fig. 5-9 The complete axis movement starts at the spindle gage line and includes the clearance, if necessary. (*Cincinnati Milacron, Inc.*)

height or distance from this surface. This surface setting is referred to as R 000 000, or the reference dimension, and all programmed depths for cutting tools and surfaces to be machined are taken from the R surface.

The R work plane (R 000 000) is generally established at 0.100 in. above the highest surface of the workpiece (Fig. 5-10). This surface is also known as *gage*

Fig. 5-10 Establishing the gage height or R 000 000 work surface location.

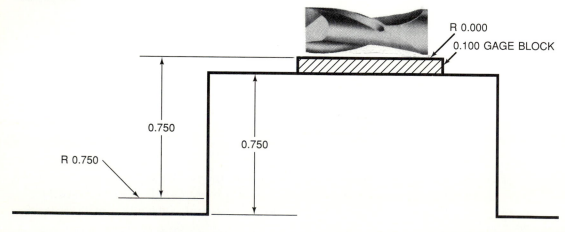

height. When cutting tools are set up, the operator generally places a 0.100-in.-thick gage on top of the highest surface of the workpiece, and all tools (regardless of length) are set to this height. Once the gage height has been set, it is not generally necessary to add this distance (0.100 in.) when changing work surfaces, since most MCUs automatically add the 0.100 in. to all future depth dimensions.

ZERO SUPPRESSION

Some machine actuation registers accept word address information in a right-to-left sequence with the decimal point fixed at four places to the left. There are seven digit positions available, so that dimensions as large as 999.9999 inches can be programmed. If a dimension of X0068750 was programmed, it would fill the register as shown in Fig. 5-11.

Some controls allow all *leading zeros*—those *before* the whole number(s) to the left of the decimal point—to be omitted. The words enter the register from right to left, and therefore the two leading zeros before the number 6 have no value and can be suppressed or omitted. This is called *leading zero suppression*. The dimension in Fig. 5-11 would be entered as X6.8750.

Other controls allow all *trailing zeros*—those *after* the number(s) to the right of the decimal point—to be omitted. The words enter the register from left to right. Therefore the one trailing zero after the number 5 has no value and can be suppressed or omitted. The elimination of insignificant zeros is called *trailing zero suppression*. The dimension in Fig. 5-11 would be entered as X006.875.

Many newer controls allow the addition of the decimal point when dimensions are entered, and on such controls both the leading and trailing zeros can be omitted. The dimension in Fig. 5-11 would be entered as X6.875.

Fig. 5-11　The location of the decimal point and digits in a standard register.

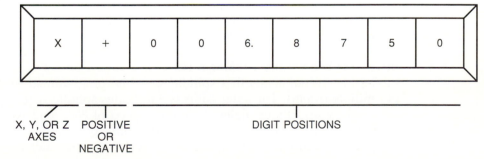

FEEDS AND SPEEDS

Feed

Feed is the amount that a cutting tool advances into the work, which generally controls the amount and rate that metal is removed from a workpiece. It is generally measured in inches per revolution (in./r) or in inches per minute (in./min). On most NC machine tools, the feed rate is coded into the machine tool in *inches per minute* (in./min) by the manufacturer. The feed rate used will depend on the rigidity of the machine tool, the work setup, and the type of material being machined.

The EIA standard feed rate code consists of the letter F plus five digits (three to the left and two to the right of the decimal point). The numbers to the left of the decimal point represent whole inches (or millimeters), while the numbers to the right of the decimal represent fractions of an inch (or millimeter). Generally when a feed rate in inches or millimeters per minute is programmed, it is given in whole numbers (to the left of the decimal). A feed rate per revolution is generally given in decimals (to the right of the decimal point). Feed can be programmed in inches (or millimeters) *per minute* (F25.5 would be a feed rate of 25½ in./min). (F0.01 would be a feed rate of 0.010 in./r.) The control unit on each machine tool will govern whether feed is programmed in inches per minute or inches per revolution.

Speed

The *speed* of a machine tool spindle generally means the number of revolutions that the spindle makes in one minute of operation. The spindle speed rate (r/min) is generally governed by the work or cutter diameter, the type of cutting tool used, and the type of material being cut. Too fast a speed rate will cause the cutting tool to break down quickly, resulting in time wasted replacing or sharpening the cutting tool. Too slow a speed rate will result in the loss of valuable time, resulting in a higher cost for each part machined. Therefore the speed rate is a very important factor which affects the production rate and also the life of the cutting tool. On NC machine control units, various methods are used to set the spindle speed. The most common are by revolutions per minute (r/min), surface feet (or meters) per minute (sf/min or sm/min) by the G96 function code which provides constant surface speed (CSS), or a three-digit code number ("Magic Three").

The EIA recommends that spindle speed be programmed in revolutions per minute (r/min). The letter address S indicates spindle speed and may be followed by up to four digits. A spindle speed of 300 r/min would be programmed as S300; a speed of 2100 r/min would be programmed as S2100.

Spindle speed may be programmed in surface feet (or meters) per minute through the G96 preparatory function code. Some MCUs (especially on turning centers) have the capabilities of maintaining CSS at the point of the cutting tool. The proper value of the surface speed is programmed under an appropriate G function code, and as a diameter changes during a machining operation, the spindle speed will automatically increase, decrease, or remain unchanged.

On some machine tools, spindle speeds can be programmed by using a three-digit coded number system, commonly called the "Magic Three." The second and third number of the three-digit code is the speed rounded off to two-digit accuracy. The first digit in the code always has a place value of 3 plus the number of digits to the left of the decimal point. For example, a speed of S610 would be as follows:

6 means	000.000	(three numbers to the left and three numbers to the right of the decimal point)
10 means	100.000	(100 r/min)

The digits are always preceded by the letter address S; see the following example standard spindle speeds.

Code	r/min
S570	70
S610	100
S614	140
S615	150
S620	200
S622	220
S630	300
S635	350
S643	430
S649	490
S669	690
S675	750
S698	980
S711	1100
S715	1500
S721	2100

MISCELLANEOUS FUNCTIONS

Miscellaneous NC functions perform a variety of auxiliary commands, such as stopping the program, starting or stopping the spindle or feed, tool changes, coolant flow, etc., which control the machine tool. They are generally multicharacter ON/OFF codes which select a function controlling the machine tool. Miscellaneous functions are used at the beginning or end of a cycle and are identified by the letter address M followed by a two-digit number.

In most cases, miscellaneous codes such as M00, M01, M02, M06, or M26 are effective only in the specific block in which they are programmed. If they are needed in two successive blocks, they must be programmed in each block. Most other miscellaneous codes do not have to be repeated in succeeding blocks.

Both preparatory and miscellaneous functions are generally classified as either *modal* or *nonmodal*. When *modal miscellaneous functions* such as M03 (spindle CW) and preparatory functions such as G81 (canned drill cycle) are programmed, they stay in effect in succeeding blocks until they are replaced by another function code. The modal function or code is changed or canceled as soon as a new miscellaneous or preparatory function code is programmed.

All *nonmodal functions* such as M00, M01, M02, M06, etc., are valid or operational only in the block programmed. If they are needed in successive blocks, they must be programmed again.

The following miscellaneous functions have been supplied by the EIA and are according to their standard E1A-274-D.

Number	Operation	Definition
M00	Program Stop	A miscellaneous function command to cancel the spindle and coolant functions and terminate further program execution after completion of other commands in the block.
M01	Optional (Planned) Stop	A miscellaneous function command similar to a program stop except that the control ignores the command unless the operator has previously validated the command.
M02	End of Program	A miscellaneous function indicating completion of workpiece. Stops spindle, coolant, and

Number	Operation	Definition
		feed after completion of all commands in the block. Used to reset control and/or machine. Resetting control may include rewind of tape to the end of record character or progressing a loop tape through the splicing leader.*
M03	Spindle CW	Start spindle rotation to advance a right-handed screw into the workpiece.
M04	Spindle CCW	Start spindle rotation to retract a right-handed screw from the workpiece.
M05	Spindle Off	Stop spindle in normal, most efficient manner; brake, if available, applied; coolant turned off.
M06	Tool Change	Stops spindle and coolant, and retracts tool to full retract position. Should be coded in last block of information in which a given tool is used.
M07– M08– M09	Coolant, On, Off	Mist (No. 2), flood (No. 1), tapping coolant or dust collector.*
M10–M11	Clamp, Unclamp	Can pertain to machine slides, workpiece, fixtures, spindle, etc.
M12	Synchronization Code	An inhibiting code used for synchronization of multiple sets of axes.
M15–M16	Motion +, Motion –	Rapid Traverse or feed direction selection, where required.*
M19	Oriented Spindle Stop	A miscellaneous function which causes the spindle to stop at a predetermined or programmed angular position.*

*The choice for a particular case must be defined in the Format Classification Sheet.
Note: Additional commands may be required on specific machines. Unassigned code numbers should be used for these and specified on the Format Classification Sheet. On certain machines the functions described may not be completely applicable; deviations and interpretations should be clarified in the Format Classification Sheet.

Number	Operation	Definition
M26	Pseudo Tool Change	Retracts tool from gage height to tool change position. Used primarily to avoid clamps or part obstructions.
M30	End of Data	A miscellaneous function which stops spindle, coolant, and feed after completion of all commands in the block. Used to reset control and/or machine. Resetting control will include rewind of tape to the end of record character, progressing a loop tape through the splicing leader, or transferring to a second tape reader.*
M31	Interlock By-Pass	A command to temporarily circumvent a normally provided interlock.*
M47	Return to Program Start	A miscellaneous function which continues program execution from the start of program, unless inhibited by an interlock signal.
M49	Override By-Pass	A function which deactivates a manual spindle or feed override and returns the parameter to the programmed value. Canceled by M48.
M59	CSS By-Pass Updating	A function which holds the RPM constant at its value when M59 is initiated/canceled by M58.
M90–M99	Reserved for User	Miscellaneous function outputs which are reserved exclusively for the machine user.

*The choice for a particular case must be defined in the Format Classification Sheet.
Note: Additional commands may be required on specific machines. Unassigned code numbers should be used for these and specified on the Format Classification Sheet. On certain machines the functions described may not be completely applicable; deviations and interpretations should be clarified in the Format Classification Sheet.

TAPE FORMAT

The format on NC punched tape refers to the sequence and arrangement of the coded information which must conform to EIA standards. Each word address under the EIA standard should be listed in its proper place in the order of words. The following word addresses used in some controllers are used as an example to show their place in the order of words. There may be variations of this order of words with different MCU manufacturers.

Character	Purpose
N	Sequence number for data block
G	Preparatory function
X	Amount of X axis travel (in. or mm)
Y	Amount of Y axis travel (in. or mm)
Z	Amount of Z axis travel (in. or mm)
R	Clearance plane for fixed cycles (in. or mm)
I	Arc center coordinate parallel to X axis (in. or mm)
J	Arc center coordinate parallel to Y axis (in. or mm)
K	Arc center coordinate parallel to Z axis or depth increment for G83 cycle (in. or mm)
F	Feed rate or dwell
S	Spindle speed
T	Tool number
H	Tool length compensation
D	Cutter radius compensation
E	Fixture offset
L	Dwell time for fixed cycles
M	Miscellaneous function
Q	Subroutine (macro)

A block of information contains five or more words, and the most common tape format uses either the *word address* or the *interchangeable* format. Figure 5-12 shows one full block of information containing 10 words and the tape coding for each specific item of information to be used on an NC turret-drilling machine.

ACRAMATIC 330 (ADVANCED)
TAPE FORMAT FOR CINTIMATIC TURRET DRILL
(WORD ADDRESS)

EXAMPLE - ONE FULL BLOCK

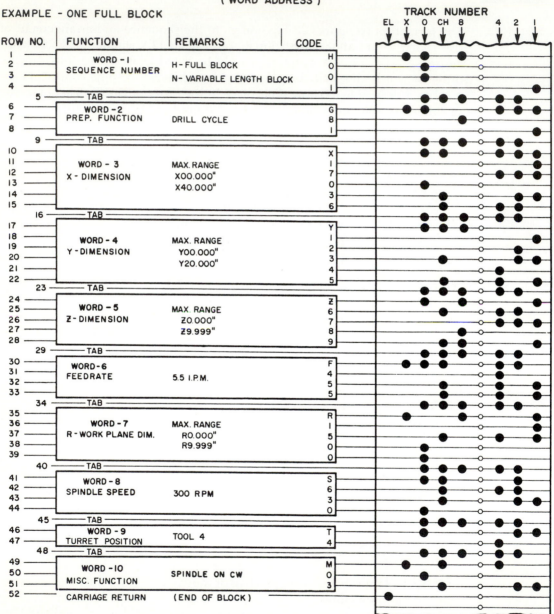

Fig. 5-12   One full block of information containing 10 words which are programmed according to the EIA standard coding. (*Cincinnati Milacron, Inc.*)

Sequence Number Coding (Fig. 5-12)

Row	Description
1	The *word address* for the sequence number which follows. An H is used for all full or alignment blocks, and an N for all others.
2–4	The *sequence number* (000 to 999) which identifies each block of information. The sequence number must always contain three digits; zeros must be included where required.
5	TAB

Preparatory Function (Fig. 5-12)

Row	Description
6	A letter G, which is the word address for the preparatory function or cycle type.
7–8	Selects the cycle type according to the following codes:

Code	Cycle
G79	Mill
G80	Cancel
G81	Drill
G82	Dwell
G84	Tap
G85	Bore

Row	
9	TAB

Coordinate Information (Fig. 5-12)

Row	Description
10	The word address letter X.
11–15	The X axis position in inches. Row 11—the tens-of-inches digit Row 12—the units-of-inches digit Row 13—the tenths-of-inches digit Row 14—the hundredths-of-inches digit Row 15—the thousandths-of-inches digit

Row	Description
	The program must contain all five digits plus the leading and trailing zeros.
16	TAB
17	The word address letter Y.
18–22	The Y axis position in inches.
	Rows 18 to 22 are the same as rows 11–15.
	Note: It is assumed that the X and Y dimensions are positive in the first quadrant (see Fig. 5-13) with the zero point placed in the lower left-hand corner of the machine table.
23	TAB
24	The word address letter Z.
25–28	The Z feed dimension in inches.
	The depth is measured and programmed from the reference work surface and does not have to be referenced to the machine table, upper limit of turret or spindle, or the gage height.
	Rows 25 to 28 are the inches, tenths-of-inches, hundredths-of-inches, and thousandths-of-inches digits, respectively.
29	TAB

Fig. 5-13 The XY zero is located in the lower left-hand corner of the machine table. (*Cincinnati Milacron, Inc.*)

PROGRAMMABLE RANGE

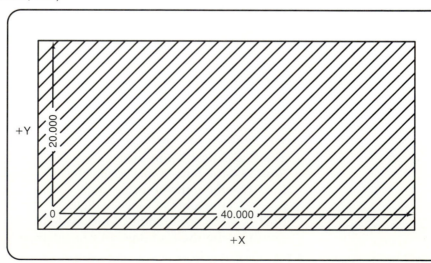

Feed Rate (Fig. 5-12)

Row	*Description*
30	The feed word address letter F.
31–33	The feed rate for the X, Y, and Z axes in inches per minute.
34	TAB

Work Plane—Z Rapid Position (Fig. 5-12)

Row	*Description*
35	The Z rapid position word address letter R.
36–39	The Z rapid position in inches (36), tenths (37), hundredths (38), and thousandths (39) of an inch.
40	TAB

Spindle Speed (Fig. 5-12)

Row	*Description*
41	The spindle speed word address letter S.
42–44	The "Magic 3" coding for spindle speeds.
45	TAB

Turret Position (Fig. 5-12)

Row	*Description*
46	The turret position word address letter T.
47	The turret position per the following codes:

Code	*Turret Position*
1	No. 1 spindle
2	
3	
4	
5	
6	
7	
8	No. 8 spindle

48	TAB

Miscellaneous Functions

Row	Description
49	The miscellaneous function word address letter M.
50–51	The miscellaneous function codes per the following:

Code	Description
M00	Program stop (spindle and coolant off)
M02	End of program
M03	Spindle on clockwise
M04	Spindle on counterclockwise
M05	Spindle off
M06	Tool change
M09	Coolant off
M13	Spindle clockwise and coolant on
M14	Spindle counterclockwise and coolant on

53	End of Block

REVIEW QUESTIONS

1. Explain the function of an NC programmer.

2. List the requirements of a successful programmer.

NC Functions

3. What two associations have established standards for NC functions?

4. Name three types of function codes used for NC programming.

Sequence Numbers

5. Explain the importance of the sequence number.

6. How are the word address codes N, H, and O used?

7. What is the purpose of assigning sequence numbers in progression of 10?

Preparatory Functions

8. Name six operations which are activated by preparatory functions.

9. Define fixed or canned cycles.

10. Why are canned cycles important in NC work?

11. Explain what happens when a G81 drill cycle is programmed.

12. What types of codes are G82, G84, and G85?

13. Define the following: G00, G01, G70, G71, G90, G91, G96, and G97.

Machine Motion Axes

14. How many motions does a single-spindle NC drilling machine have?

15. Define each axis of an NC drilling machine.

16. How can any point be positioned accurately on a flat plane?

17. Define "origin" or "zero point."

18. With a suitable illustration explain X and Y positive and negative dimensions.

19. In what sequence should axis motion be programmed if more than one axis is required?

20. Describe the Z axis and explain the difference between positive and negative movement.

21. What two motions occur when the Z axis is activated?

22. What does the word address letter R represent?

23. Define *work plane* and *gage height*.

Zero Suppression

24. Explain leading and following zeros.

25. What is meant by *zero suppression?*

26. When can both leading and trailing zeros be omitted in NC programming?

Feeds and Speeds

27. Define *feed.*

28. What factors affect the feed rate used during machining?

29. What word address letter is used to program feed rate?

30. Define *speed*.

31. What factors affect the speed used during a machining operation?

32. Explain the effect of a speed rate which is:
   (a) Too slow
   (b) Too fast

Miscellaneous Functions

33. Name four auxiliary commands which miscellaneous functions perform.

34. Explain the difference between modal and nonmodal functions.

35. Define the following miscellaneous functions:
   (a) M00
   (b) M02
   (c) M03
   (d) M06
   (e) M30

Tape Format

36. Define each word in the following block of information:

N20 G83 X1.5 Y−0.5 Z−1.5 M03

CHAPTER

SIX

Simple Programming

Simple programming consists of taking information from a part drawing and converting this information into a language that can be put on punched tape or sent directly to the machine tool for it to perform one or a series of operations. This information is generally contained on paper tape which has been prepared by someone who is familiar with the language that the computer numerical control (CNC) machine understands. This language (called a *machine language*) consists of numbers, letters, symbols, etc.

The person who can take a part drawing (or print) of a workpiece, determine what sequence of machining operations are required, and prepare a numerical control (NC) tape so that the controls of a machine tool will reproduce the part is called a *part programmer* (Fig. 6-1). To properly program any CNC machine, the part programmer must be familiar with the coding and abbreviations that are common to NC and the machine control unit (MCU) of that particular machine. The part programmer must also be familiar with the proper machining procedures and with the tools required for each operation, and be able to communicate this information accurately to the controls that operate the machine tool.

After completing
this chapter,
you should be
able to:

1. Understand the codes and functions required for NC programming

2. Prepare an NC manuscript from information supplied on a print

3. Write a simple program for machining a part

POINT-TO-POINT OR CONTINUOUS PATH

NC programming falls into two distinct categories (Fig. 6-2).

The difference between the two categories was once very distinct. Now, however, most control units are able to handle both point-to-point and continuous path machining. A knowledge of both programming methods is necessary to understand what applications each has in NC.

POINT-TO-POINT POSITIONING

Point-to-point positioning is used when it is necessary to accurately locate the spindle, or the workpiece mounted on the machine table, at one or more specific locations to perform such operations as drilling, reaming, boring, tapping, and punching (Fig. 6-3). Point-to-point positioning is the process of positioning from one coordinate (XY) position or location to another, performing the machining operation, and continuing this pattern until all the operations have been completed at all programmed locations.

Drilling machines or point-to-point machines are ideally suited for positioning the machine tool (say a drill) to an exact location or point, performing the machining operation (such as drilling a hole), and then moving to the next

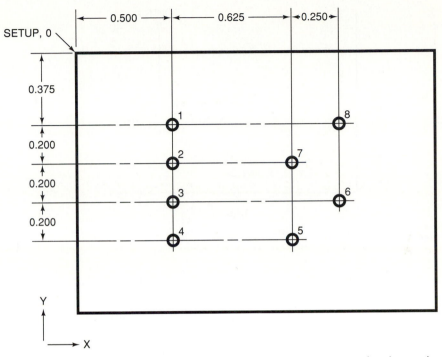

The programmer takes information and specifications from a part drawing and records the machining sequence, tools required, etc., on a program sheet or manuscript. (*The Superior Electric Company*)

location (where another hole could be drilled). As long as each point or hole location in the program is identified, this operation can be repeated as many times as required.

Point-to-point machining moves from one point to another as fast as possible ("rapids") while the cutting tool is above the work surface. "Rapid travel"

Fig. 6-2 Types of NC positioning systems.

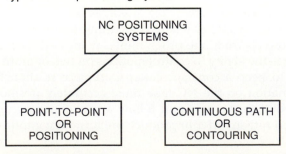

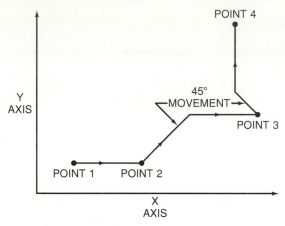

Fig. 6-3 The path followed by point-to-point positioning to reach various programmed points (machining locations) on the XY axis.

is used to quickly position the cutting tool or workpiece between each location point before a cutting action is started. The rate of rapid travel is usually between 150 and 400 in./min (38 and 101 m/min). Both XY axes move simultaneously and at the same rate during rapid traverse. This results in a movement along a 45° angle line until one axis is reached, and then there is a straight line movement to the other axis.

In Fig. 6-3, point 1 to point 2 is a straight line, and the machine moves only along the X axis; but points 2 and 3 require that motion along both the X and the Y axes take place. As the distance in the X direction is greater than in the Y direction, Y will reach its position first, leaving X to travel in a straight line for the remaining distance. A similar motion takes place between points 3 and 4.

CONTINUOUS PATH (CONTOURING)

Contouring, or *continuous path machining,* involves work such as that produced on a lathe or milling machine, where the cutting tool is in contact with the workpiece as it travels from one programmed point to the next. Continuous path positioning is the ability to control motions on two or more machine axes simultaneously to keep a constant cutter-workpiece relationship. The programmed information on the NC tape must accurately position the cutting tool from one point to the next and follow a predefined accurate path at a programmed feed rate in order to produce the form or contour required (Fig. 6-4).

The method by which contouring machine tools move from one pro-grammed point to the next is called *interpolation*. This ability to merge individual axis points into a predefined tool path is built into most of today's MCUs. There are five methods of interpolation: linear, circular, helical, parabolic, and cubic. All contouring controls provide linear interpolation, and most controls are capable of both linear and circular interpolation. Helical, parabolic, and cubic interpolation are used by industries that manufacture parts which have complex shapes, such as aerospace parts and dies for car bodies.

Linear Interpolation

Linear interpolation consists of any programmed points linked together by straight lines, whether the points are close together or far apart (Fig. 6-5). Curves can be produced with linear interpolation by breaking them into short, straight-line segments. This method has limitations, because a very

Fig. 6-4   Types of contour machining. (A) Simple contour; (B) complex contour. (*Allen-Bradley*)

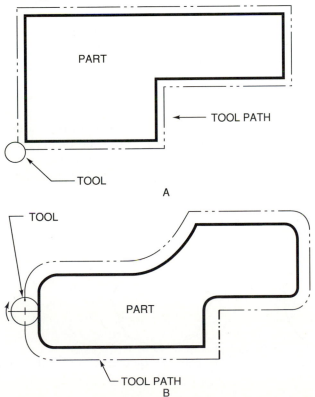

PART

TOOL PATH

TOOL

A

TOOL

PART

TOOL PATH

B

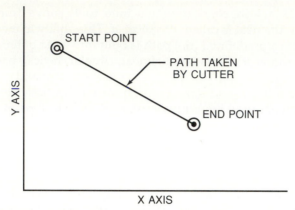

Fig. 6-5
An example of two-axis linear interpolation.

large number of points would have to be programmed to describe the curve in order to produce a contour shape.

A contour programmed in linear interpolation requires the coordinate positions (XY positions in two-axis work) for the start and finish of each line or segment. Therefore, the end point of one line or segment becomes the start point for the next segment, and so on, throughout the entire program.

The accuracy of a circle or contour shape depends on the distance between each two programmed points. If the programmed points (XY) are very close together, an accurate form will be produced. To understand how a circle can be produced by linear interpolation, refer to Fig. 6-6. Figure 6-6B shows that when 8 connecting lines are machined, an octagon is produced, while Fig. 6-6C shows 16 connecting or chord lines. If the chords in each are examined, it should be apparent that the more program points there are, the more closely the form resembles a circle. If there were 120 chord lines, it would be

Fig. 6-6   Linear interpolation with arcs and circles. (A) True circle; (B) eight-segment circle; (C) sixteen-segment circle.

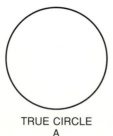

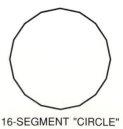

TRUE CIRCLE   8-SEGMENT "CIRCLE"   16-SEGMENT "CIRCLE"
A   B   C

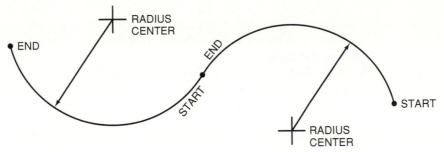

Fig. 6-7 Any complex form on two axes can be generated by circular interpolation.

difficult to see that the circle was made up of a series of short chord lines. Therefore, on a control unit which is capable only of linear interpolation, very accurate contours require very long programs because of the large number of points which need to be programmed.

Circular Interpolation

The development of MCUs capable of *circular interpolation* has greatly simplified the process of programming arcs and circles. To program an arc (Fig. 6-7), the MCU requires only the coordinate positions (the XY axes) of the circle center, the radius of the circle, the start point and end point of the arc being cut, and the direction in which the arc is to be cut (clockwise or counterclockwise). See Fig. 6-8. The information required may vary with different MCUs.

The circular interpolator in the MCU breaks up the distance of each chord line (circular span) into a series of the smallest movement increments caused

Fig. 6-8 For two-dimensional circular interpolation the MCU must be supplied with the XY axis, radius, start point, end point, and direction of cut.

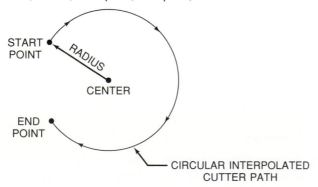

by one single output pulse. This is usually 0.0001 in. (0.002 mm), and the interpolator automatically computes enough output pulses to describe the circular form and then provides the control data required for the cutting tool to produce the form.

The advantage and power of circular interpolation are better appreciated by comparing it to linear interpolation. More than a thousand blocks of information would be needed to produce a circle in linear interpolation segment by segment. Circular interpolation requires only five blocks to produce the same circle.

Helical Interpolation

Helical interpolation combines two-axis circular interpolation with a simultaneous linear movement in the third axis. All three axes move at the same time to produce the helical (spiral) path required. Helical interpolation is most commonly used in milling large internal-diameter threads or helical forms.

Parabolic Interpolation

Parabolic interpolation has its greatest application in automotive dies, mold work, and any form of sculpturing. Parabolic interpolation is a method of creating a cutter path covering a wide variety of geometric shapes, such as circles, ellipses, parabolas, and hyperbolas. Parabolic interpolation can be defined as a movement that is either totally parabolic or part parabolic and has three non-straight-line locations (two end points and one midpoint). This allows it to closely simulate curved sections, using about 50 times fewer program points than would be required using linear interpolation. The more powerful computers of today have put parabolic interpolation in the background because it is now possible to use simpler interpolation.

Cubic Interpolation

Cubic interpolation enables sophisticated cutter paths to be generated in order to cut the shapes of the forming dies used in the automobile industry. These dies can be machined with only a small number of input data points in the program. Besides describing the geometry, cubic interpolation smoothly blends one curved segment to the next without the interruption of boundary (start and stop) points. However, a larger-than-normal computer memory is required to handle the extensive program for cubic interpolation.

Interpolation Summary

The function of interpolation is to store programmed information, monitor and direct the machine axis motions, and keep a straight-line motion between coordinate points at a defined vector feed rate. The purpose of nonlinear interpolation is to eliminate the need to program numerous coordinate points for curved surfaces, which would be necessary with linear interpolation. The control units of today have the mathematical ability to generate the numerous points needed to produce curved surfaces if they are provided with a description of the curve.

MANUSCRIPT

Before a program for any workpiece for a CNC machine tool is started, the job print should be studied carefully. It should first be noted which surfaces of the workpiece must be machined, what special operations are required, and the dimensional tolerances which are required, in order to determine the sequence of machining operations. It is wise to remember that the machining of a part, whether by conventional machining or by NC machining, is basically the same. In conventional machining, a skilled operator moves the machine slides manually, while in NC machining, the machine slides are moved automatically from the information supplied by the NC tape or by a computer.

Manuscript Data Information

It is the job of the programmer to see that the machine tool receives the proper information in order to cut the part to the proper shape and size. Using numerical language, the programmer must record on a prepared form (called a *manuscript*) all the instructions that the machine tool must have in order to complete the job. The manuscript should contain all the machine tool movements, cutting tools required, speeds, feeds, and any other information which might be required to machine the part (Fig. 6-9). This information should be in a uniform format and be as clear as possible to give the NC machine operator a good understanding of what is required. Figure 6-10 shows the types of information which should be included on or with a manuscript and should be supplied by the programmer.

1. *Part Sketch*
   A rough sketch should be made of the part, and incremental or absolute dimensions should be given for each axis location from the zero or reference point.

2. *Zero (or Reference) Point*
   - A zero or reference point should be established to permit the alignment of the workpiece and the machine tool.

PREPARED BY *LGH*		PART NAME			PART NO.		OPER. NO.
DATE *3-27-67*			*PLATE*		*BG 75507*		*—*
CK'D BY *REC*							
DATE *3-28-67*		REMARKS:					
SHEET *1* OF *1*		*RUN PROGRAM TWICE, ONCE WITH*					
DEPT *16*		*CENTER DRILL, ONCE WITH 3/8 DRILL*					
TAPE NO *127*		*SET FEED AT HI, TOOL AT AUTO, BACKLASH AT #2*					

SEQ NO	TAB OR EOB	OR	"X" INCREMENT	TAB OR EOB	OR	"Y" INCREMENT	TAB OR EOB	"M" FUNCT	EOB	INSTRUCTIONS
	EOB									
RWS	*EOB*									*CHANGE TOOL, LOAD, START*
1	*TAB*		*2000*	*TAB*	*–*	*1875*	*EOB*			
2	*TAB*		*2000*	*EOB*						
3	*TAB*		*2000*	*EOB*						
4	*TAB*			*TAB*	*–*	*1875*	*EOB*			
5	*TAB*	*–*	*2000*	*EOB*						
6	*TAB*	*–*	*2000*	*EOB*						
7	*TAB*	*–*	*2000*	*TAB*		*3750*	*TAB*	*02*	*EOB*	

Fig. 6-9 All the information necessary to machine a part to size and shape must be included on a manuscript. (*The Superior Electric Company*)

- A tool change position should also be selected which will provide enough room to change cutting tools and load and unload parts.

3. *Workholding Device*
   - The device or fixture best suited to hold the part securely and not interfere with the machining operations should be selected.
   - The fixture should not have any of its components too high above the part to be machined.
   - The setup instructions for the fixture should be included on the manuscript.

4. *Sequence of Operations*
   - Select what operations should be done first and in what sequence so that the part is machined to size and shape in the shortest time possible.
   - A knowledge of basic machining operations is essential to list the operation sequence properly.

5. *Axis Dimensions*
   - All the data necessary for every movement of the table or the cutting tool must be listed.
   - This must include axes locations for every surface to be machined or every hole to be drilled, tapped, reamed, etc.

6. *Tool List and Identification*
   - Whenever tools are required, they should be indicated in the Remarks or Comment column of the manuscript.
   - The tool identification number should indicate the order in which it will be used to machine the part, the tool diameter and length, and the tool and offset number.

Fig. 6-10   The points which must be considered when preparing a manuscript to ensure good communications between the programmer and the machine tool operator.

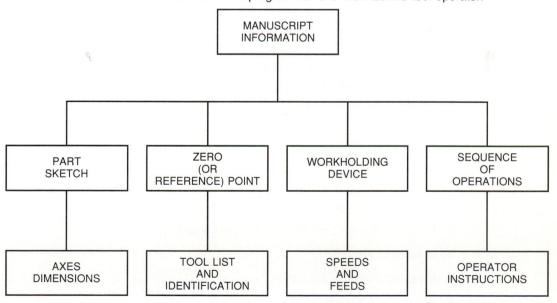

7. *Speeds and Feeds*

- The speed in revolutions per minute (r/min) that each cutting tool must rotate for the most efficient machining of the part must be listed on the manuscript.
- The feed rate in inches or meters per minute (in./min or m/min) that will give the best cutting tool life while maintaining efficient machining rates should be listed for each cutter.

8. *Operator Instructions*

- Special instructions to the operator should be included in the Remarks or Comments column.
- These should include such information as specifications about cutting tools, loading or unloading the workpiece, time to change cutting tools if it is not an NC function, etc.

PROGRAMMING PROCEDURE

The part illustrated in Fig. 6-11 will be used to introduce simple programming in easy-to-understand steps. Each step in the programming procedure will be explained in detail in order to provide the reader with a clear understanding of the meaning of the various codes, axis movements, etc., and what happens as a result of each programming step.

Two programs (one in incremental and the other in the absolute mode) will be written, first, to trace the part boundary and, second, to locate the position of the holes. Another program will be written in both the incremental and the absolute mode to drill the six ⅜-in.- (9.5-mm-) diameter holes. In practice, the programming for hole locations and the drilling operation would be combined in the same program.

Notes

1. All programming begins at the zero or reference point (XY zero), which is located at the left of the part. This allows clearance for changing cutting tools and loading or unloading parts.

2. Program the part boundary *clockwise,* starting at point A.

3. Return to XY zero.

4. Program the hole locations.

5. Return to XY zero.

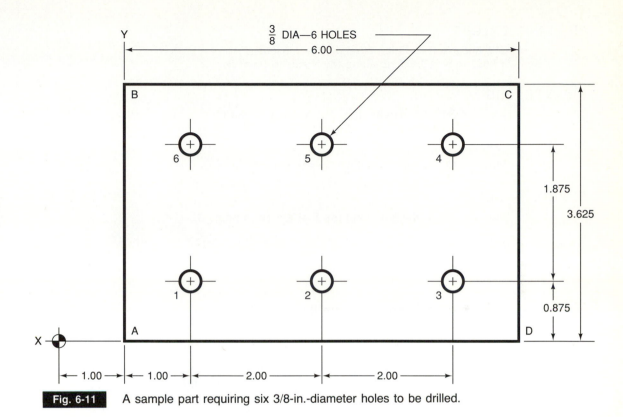

Fig. 6-11 A sample part requiring six 3/8-in.-diameter holes to be drilled.

The Part Boundary Program (Incremental)

%

Rewind stop code/parity check.

N010 G91

Incremental program mode.
Sequence numbers should be in progressions of 10 to leave room for other steps to be inserted if necessary.

N020 G70

Inch mode.

N030 G00 X1.000

G00 rapid traverse mode.
X1.000 the cutting tool is located 1.000 to the right along the X axis to point A.

`N040 G01 Y3.625`

> G01 linear interpolation (straight-line movement).
> Y3.625 moves 3.625 up along the Y axis to point B.

`N050 X6.000`

> Moves 6.000 to the right along the X axis to point C.

`N060 Y-3.625`

> Moves 3.625 down along the Y axis to point D.

`N070 X-6.000`

> Moves 6.000 to the left along the X axis back to point A.

`N080 G00 X-1.000`

> Returns to XY zero in the rapid mode.

The Hole Locations Program (Incremental)

`N090 X2.000 Y0.875`

> Rapids 2.000 along the X axis and 0.875 up the Y axis to hole location #1 because it is still in the rapid mode from sequence number N080.

`N100 X2.000`

> Rapids 2.000 along the X axis to hole #2.

`N110 X2.000`

> Rapids to hole #3.

`N120 Y1.875`

> Rapids 1.875 up the Y axis to hole #4.

`N130 X-2.000`

> Rapids to the left 2.000 to hole #5.

`N140 X-2.000`

> Rapids to the left to hole #6.

`N150 X-2.000 Y-2.750`

> Rapids back to XY zero.

`N160 M30`

> End of program.

`%`

> Rewind/stop code.

The Part Boundary Program (Absolute)

`%`

> Rewind stop code/parity check

`N010 G90`

> Absolute program mode
> Sequence numbers should be in progressions of 10 to leave room for other steps to be inserted if necessary.

`N020 G70`

> Inch mode.

`N030 G00 X1.000`

> G00 rapid traverse rate.
> X1.000 the cutting tool is located 1.000 to the right along the X axis to point A.

`N040 G01 Y3.625`

> G01 linear interpolation (straight-line movement)
> Y3.625 moves 3.625 up along the Y axis to point B.

`N050 X7.000`

> Moves 6.000 to the right along the X axis to point C (a 6.000 move is 7.000 from the XY zero).

`N060 Y0.0`

> Moves down 3.625 along the Y axis to point D.

`N070 X1.000`

> Moves 6.000 to the left along the X axis back to point A.

`N080 X0.0`

> Returns to the XY zero in the rapid mode.

The Hole Locations Program (Absolute)

N090 X2.000 Y0.875

Rapids 2.000 along the X axis and 0.875 up the Y axis to hole location #1 because it is still in the rapid mode from sequence number N080.

N100 X4.000

Rapids 2.000 along X axis to hole #2 (this is 4.000 to the right of the XY zero).

N110 X6.000

Rapids 2.000 (6.000 from XY zero) along the X axis to hole #3.

N120 Y2.750

Rapids 1.875 (2.750 from XY zero) up the Y axis to hole #4.

N130 X4.000

Rapids 2.000 to the left to hole #5.

N140 X2.000

Rapids 2.000 to the left to hole #6

N150 X0.0 Y0.0

Rapids back to XY zero.

N160 M30

End of program.

%

Rewind/stop code.

Program #2—Drilling Holes

A review of the R work plane, the Z axis motion, and fixed or canned cycles covered in Chapter 5 helps make clear how the holes for the workpiece shown in Fig. 6-11 can be drilled as efficiently as possible.

- The *R work plane*, or gage height, is generally 0.100 above the surface of a workpiece and is used as a reference, and all other work surfaces are relative to this location.
- The *Z axis motion* moves the cutting tool either into the workpiece (a minus motion) or away from the workpiece (a plus motion).

- *Fixed or canned cycles* (G81 to G89) are preset combinations of operations, such as drilling, where all machine axes motions are programmed and will repeat themselves until canceled by a G80 code.

As can be seen from the following example, all these factors can be programmed and will repeat themselves until they are canceled by the G80 code.

```
N040 G81 X2.000 Y1.500 R0.100 Z-1.000 F5
```

```
G81
```

A fixed or canned drilling cycle.

```
R0.100
```

The gage height is set at 0.100 above the work surface.

```
Z-1.000
```

The drill will be fed into the work 1.000 deep.

```
F5
```

The feed rate for the drill will be 5 in./min.

After reaching the Z depth, the drill will automatically retract in the rapid mode to the gage height.

Program #2—Incremental

To drill the six ⅜-in.-diameter holes illustrated in Fig. 6-11, the following points should be kept in mind:

Notes

1. The machining operation is drilling.

2. The zero or reference point (XY zero) is located to the left of the part.

3. The G81 fixed or canned drilling cycle will be used to eliminate repetition in programming.

4. Both incremental and absolute programming will be shown.

```
N010 G91
```

Incremental mode.

N020 G70

    Inch mode.

N030 G81 X2.000 Y0.875 R0.100 Z−1.000 F5 M03

G81

    Fixed drill cycle.

X2.000 }
Y0.875 }

    The machine table will "rapid" to hole #1 position.

R0.100

    The machine spindle will rapid down so that the drill point is 0.100 in. above the surface of the part.

Z−1.000 }
F5   }

    The drill will advance 1.000 in. into the workpiece at a feed rate of 5 in./min.
    The drill will rapid out of the hole back to gage height (0.100 above work).

M03

    Turns the spindle on to revolve in a clockwise direction
    The M03 or M04 code is the only one that can be used in a fixed cycle.

N040 X2.000

    The table will rapid 2.000 in. to hole #2 position.
    The G81 fixed cycle (N030) will be repeated, and a hole will be drilled at this position.

N050 X2.000

    The table will rapid 2.000 in. to hole #3 position.
    The G81 cycle will be repeated, and a hole will be drilled.

N060 Y1.875

    The table will rapid 1.875 to hole #4 position. The G81 cycle will repeat.

N070 X−2.000

    The table will rapid −2.000 to hole #5 position.
    The G81 cycle will repeat.

N080 X-2.000

> The table will rapid −2.000 to hole #6 position.
> The G81 cycle will repeat.

N090 G80

> Cancels the drill cycle and automatically puts the machine in the rapid mode.

N100 X-2.000 Y-2.750 M06

> The table rapids simultaneously along the XY axes and returns to the XY zero.
> M06 stops the machine spindle and raises the cutting tool to the full retract position.

N110 M30

> Rewinds the tape in preparation for use in drilling the next part.

M00

> Optional stop, used only on a manual tool change machine.
> Safety factor to prevent machine from starting while tools are being changed.
> The cycle must be reactivated by pressing the cycle start control.

Program #2—Absolute

N010 G90

> Absolute mode.

N020 G70

> Inch mode.

N030 G81 X2.000 Y0.875 R0.100 Z-1.000 F5 M03

> The machine rapids to hole #1 location, and the spindle rapids to 0.100 above the work surface.
> The drill will be fed 1.000 in. into the part at a 5-in. feed rate and then rapid to 0.100 above the part.
> M03 turns the spindle on to revolve in a clockwise direction.

N040 X4.000

> The table rapids to hole #2 position 4.000 in. from the XY zero or reference point.
> The G81 fixed cycle (N030) will be repeated, and a hole will be drilled.

N050 X6.000

> The table will rapid to hole #3 position.
> The G81 cycle will repeat.

N060 Y2.750

The table will rapid to hole #4 position.
The G81 cycle will repeat.

N070 X4.000

The table will rapid to hole #5 position.
The G81 cycle will repeat.

N080 X2.000

The table will rapid to hole #6 position.
The G81 cycle will repeat.

N090 G80

Cancels the drill cycle and automatically puts the machine in the rapid mode.

N100 X0.00 Y0.00 M06

The table will rapid back to the XY zero or reference point.
M06 stops the machine spindle and raises the cutting tool to the full retract position.

N110 M30

Rewinds the tape in preparation for use in drilling the next part.

M00

Optional stop, used only on a manual tool change machine.
Safety factor to prevent machine from starting while tools are being changed.
The cycle must be reactivated by pressing the cycle start control.

NC MACHINING EXERCISE

Progressing from the very basic programming for the linear interpolation which was used for the part in Fig. 6-11, we shall now look at the sample part shown in Fig. 6-12. There are five stages which will be used to program this part for an NC vertical milling machine, starting with machining the edges to size and ending up with the finished part. Stages 1 and 2 will be covered later in this chapter, while stages 3, 4, and 5 (milling the triangle, circular grooves, and drilling the holes) will be covered in Chapter 7. Every stage will introduce more advanced programming techniques and the theory required.

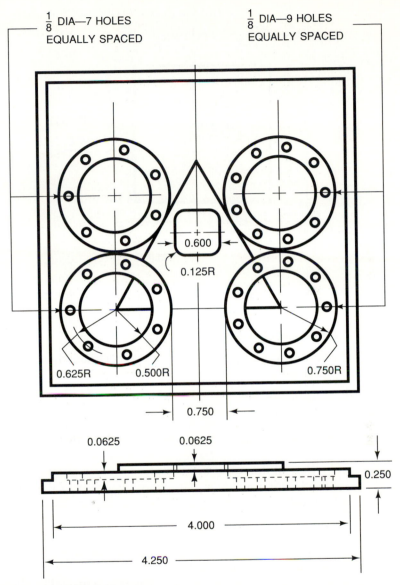

$\frac{1}{8}$ DIA—7 HOLES
EQUALLY SPACED

$\frac{1}{8}$ DIA—9 HOLES
EQUALLY SPACED

0.600

0.125R

0.625R　0.500R　0.750R

0.750

0.0625　0.0625

0.250

4.000

4.250

An NC milling and drilling exercise.

Stage 1 involves milling the four edges of part shown in Fig. 6-12 to size. Since a 1-in.-diameter milling cutter will be used for this operation, it is necessary to understand *cutter-diameter compensation* and *axis presets* in order to prepare the program required to machine the edges to the correct size.

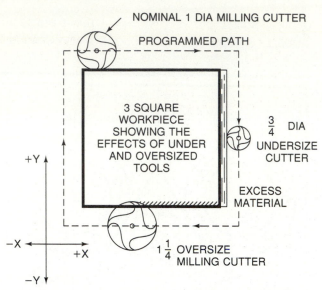

NOMINAL 1 DIA MILLING CUTTER

PROGRAMMED PATH

3 SQUARE WORKPIECE SHOWING THE EFFECTS OF UNDER AND OVERSIZED TOOLS

$\frac{3}{4}$ DIA UNDERSIZE CUTTER

EXCESS MATERIAL

+Y

−X   +X

−Y

$1\frac{1}{4}$ OVERSIZE MILLING CUTTER

Fig. 6-13   Compensation must be made for various milling cutter diameters when programming.

Cutter-Diameter Compensation

Since the basic reference point of the machine tool is never at the cutting edge of a milling cutter, the programmer must take into consideration the diameter of a cutter when programming. The center of the milling cutter is not the part that does the cutting; rather, it is some point on the periphery of the cutter. If a 1-in. end mill is used to machine the edges of a workpiece, the programmer would have to keep a ½-in. offset from the normal surface in order to cut the edges accurately (Fig. 6-13). The ½-in. offset represents the distance from the centerline of the cutter or machine spindle to the edge of the cutter which will be machining the edges. Therefore, whenever a part is being machined with some type of milling cutter, the programmer must always calculate an offset path, which is usually half the diameter of the cutter used.

On some newer MCUs, which have part surface programming, the cutter centerline offsets are calculated automatically once the diameter of the cutter for each operation is programmed. Many MCUs have operator-entry capabilities which can compensate for differences in cutter diameters; therefore an oversize cutter or one that has been sharpened can be used as long as the compensation for oversize or undersize cutters is entered.

Axis Presets

Some MCUs have two position registers, the absolute position register and the command accumulator register (Fig. 6-14). The *absolute position register* displays a continual record of the position of the machine table from the absolute zero, or home, position. This XY zero (or home) position of a machine is generally in the top right-hand corner of the table, though it can be in any corner depending on the manufacturer. The *command accumulator register* displays a continual record of the axis absolute position in relation to the programmer's (or operator's) defined XY zero or reference point.

The programmer often uses an XY zero or reference point which is at some position away from the workpiece to allow for tool changes and loading or unloading workpieces. The home position of the machine table and the XY zero of the programmer often are not the same. Most MCUs have the capability of shifting the machine zero to any point within the working range of the machine table so that it matches the programmer's XY zero. The *zero shift control* on the MCU allows the XY zero position to be shifted from the machine's absolute zero (home position) to the programmer's XY zero.

Fig. 6-14 The zero shift control of the MCU allows the XY zero location to be moved from the table's absolute (home) position to any point within the working range of the table.

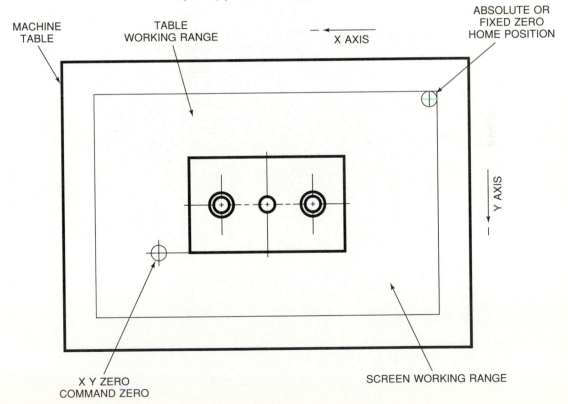

Stage 1—Machining the Edges

The edges on the part illustrated in Fig. 6-15 must be machined to 4.000 in. square. Before this can be accomplished, the following points must be considered.

Notes

1. The part must be held in a vise in order to allow the four edges to be machined in one setup.

2. A 1-in.-diameter six-flute end mill will be used for machining the edges.

3. The length (end) of the end mill must be preset to the reference plane (R) 0.100 above the part surface.

4. The zero shift control must be used to locate the programmer's XY zero (or origin).

5. The part material is aluminum.

6. The edges will be machined by climb milling. Climb milling requires less machine power and produces better accuracy and surface finish.

The program for milling the edges is as follows:

```
N010 G91
```
Incremental mode

```
N020 G70
```
Inch mode

```
N030 G00 X1.500 Y1.500 R0.100 S3000 F10 M03
N040 G01 Z-0.225
N050 Y5.000
N060 X5.000
N070 Y-5.000
N080 X-5.000
N090 G00 X-1.500 Y-1.500 M06
```

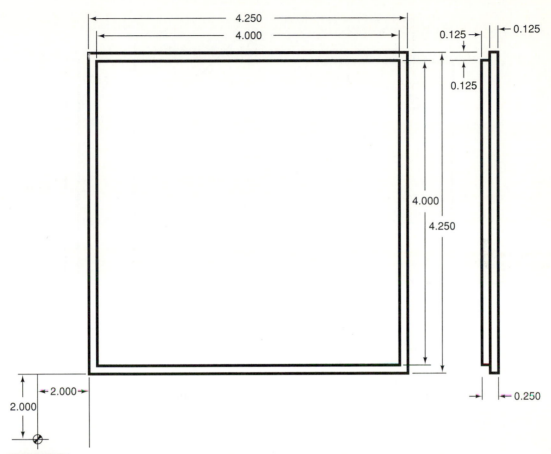

Fig. 6-15 Machining the two edges of the milling and drilling exercise to size.

An MCU equipped with *full floating zero* allows the operator to fasten a workpiece at any position on the machine table which is convenient. Once the workpiece is securely fastened, the alignment positions obtained from the part manuscript are entered into the machine controls. With the use of zero shift control or a dial, the XY zero of the manuscript is accurately located. The full floating zero provides the operator with flexibility and greatly reduces the setup time because the XY zero can be located at any place on the machine table.

Stage 2—Drilling the Holes and Machining the Recess

Stage 2 involves drilling the four ¼-in.-diameter holes in the correct locations and milling the center recess to depth and size (Fig. 6-16). Since different-length cutting tools are used for these operations, it is necessary to under-

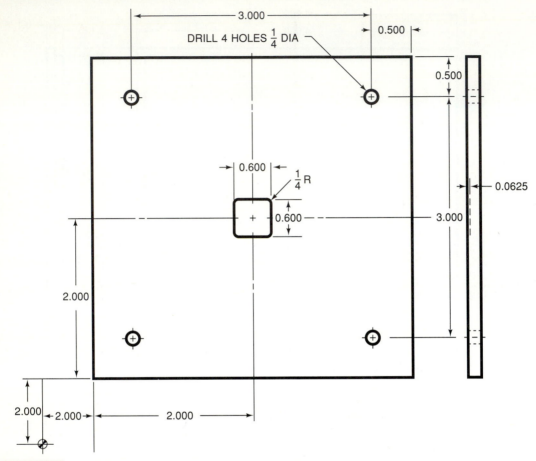

Fig. 6-16 Drilling the four 1/4-in. holes and milling the center recess.

stand *tool length compensation* so that the holes are drilled through the part but not into the vise, parallels, or any part of the workholding device. Making allowances for the lengths of various cutting tools will also ensure that the center recess will be cut to the correct depth.

Tool Length Compensation

A part programmer's initial planning must include the operation-by-operation sequence of how a part will be machined. At the same time, a list of cutting tools used for each operation must be included. This *tool list* must include the type of each cutting tool and its diameter and length. Since a wide

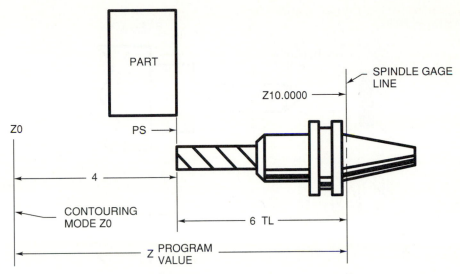

Fig. 6-17 The tool "rapids" to the gage height and then travels at a set feed rate to the Z depth.

variety of cutting tools are generally used for machining a part, some allowance must be made for the differences in their length to ensure that machining operations are performed to the correct depths.

The programmer must specify a minimum length, which is based on the maximum depth to which each cutting tool must travel. Each tool is then programmed as though it had *zero length,* and the spindle/tool gage line and the tool point were the same. The amount of Z axis travel is programmed from the R plane (0.100 above the work surface) at a specified feed rate (Fig. 6-17). The actual length of each tool is entered into the MCU either by NC tape or manually by the machine operator.

Some industries preset all cutting tools to allow for the differences in tool length, and this results in consistency of length every time a particular tool is used. Some companies use *tool assembly drawings,* which describe the cutting tool and give the setting length for each tool. Each cutting tool is assigned a specific number which is stored in the MCU and can be recalled any time that specific tool is used.

Some MCUs are equipped with *semiautomatic tool compensation* to make allowances for differences in cutting tool lengths. The tool length compensation feature provides for as many as 64 or more tool length offsets. The programmer uses the same basic tool length for all tools. When the machine is being set up, the operator mounts each tool, generally starting with the longest, sets

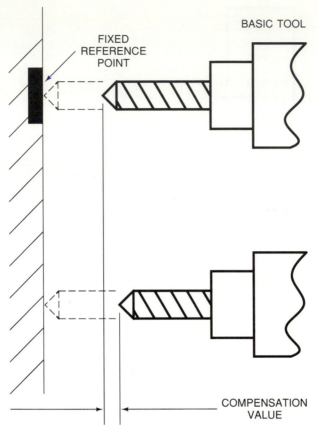

BASIC TOOL

FIXED
REFERENCE
POINT

COMPENSATION
VALUE

Fig. 6-18 Semiautomatic tool compensation automatically calculates the difference between basic tool length and actual tool length. (*Modern Machine Shop*)

it against a fixed reference point, and presses a button on the MCU. This automatically records the setting into memory. All the cutting tools for a particular job are set against the same reference point, and the difference between each tool's actual length and the basic tool length is entered into memory as a compensation value (Fig. 6-18). Individual offsets are activated or recalled from memory by H words or an M06 tool change code.

Drilling and Machining the Recess (Stage 2)

The four drilled holes and the recess shown in Fig. 6-16 must be machined. The following notes should be considered before proceeding with the machining operations.

Notes

1. A stubby ¼-in.-diameter drill must be used to produce the holes. A short stubby drill will maintain good locational accuracy and eliminate the need for center drilling all hole locations.

2. A ¼-in.-diameter two-flute end mill is to be used to mill the recess. A two-flute end mill is center-cutting and allows the end mill to plunge cut to depth.

3. The length of the drill and the end mill must be preset to the reference plane (R) 0.100 above the part surface.

   The program for *drilling the holes and machining the recess* follows Stage 1 of milling the edges of the part and is still in the incremental mode.

Drilling

```
N100 G81 X2.500 Y2.500 R0.100 Z-0.400 S3000 F5 M03
N110 Y3.000
N120 X3.000
N130 Y-3.000
N140 G80 X-5.500 Y-2.500 M06
```

The drill must be removed from the machine spindle and replaced with a ¼-in.-diameter two-flute end mill.

Milling the Recess

```
N150 G00 X4.000 Y4.000 R0.100 S3000 F1 M03
N160 G01 Z-0.1625
N170 X-0.175
N180 Y-0.175
N190 X0.350
N200 Y0.350
N210 X-0.350
N220 Y-0.350
N230 G00 Z0.1625
N240 X-3.825 Y-3.825 M06
```

The operation of milling the triangle, circular slots, and drilling holes on the bolt circle will be covered in Chapter 7.

REVIEW QUESTIONS

1. Of what does NC machine language consist?

2. Name three desirable qualities of a part programmer.

Point-to-Point or Continuous Path

3. Name the two NC positioning systems and state the purpose for which they are used.

4. What is the purpose of rapid traverse, and when is it used?

5. At what angle does rapid traverse travel when movement along the XY axis is required simultaneously?

6. What is *continuous path positioning (contouring)*?

7. Define *interpolation*.

8. For what purpose is linear interpolation used?

9. Why is linear interpolation not often used to produce accurate arcs and contours?

10. What four pieces of information are required to program an arc by circular interpolation?

11. How small is the distance of each chord line of the circular interpolator in the MCU?

12. What axis movement occurs in helical interpolation?

13. For what purpose is parabolic interpolation used?

14. Why is cubic interpolation widely used in the automotive industry?

Manuscript

15. Explain the difference between machining a part by conventional machining and by NC.

16. What information should be included on a manuscript?

17. What is the purpose of the zero or reference point?

18. Why should the information regarding the tool list and identification be included on a manuscript?

19. What is the purpose of the operator instructions?

Programming Procedure

20. Why is the XY zero or reference point usually located off the edge of a part?

21. Explain the difference between the absolute and incremental programming mode.

22. Define the following codes:
(a) G91
(b) G70
(c) G00
(d) M30

23. Use a diagram to illustrate the direction of movement of the following: X+, X−, Y+, Y−

Drilling Holes

24. Define the R work plane and the Z axis motion.

25. What are fixed or canned cycles and why are they useful in NC programming?

26. Describe what happens in a complete fixed or canned cycle.

27. What is the purpose of the following codes?
(a) G90
(b) G80
(c) M06

Cutter Diameter Compensation

28. Why is cutter diameter compensation important to programming operations involving milling cutters?

29. How do newer MCUs compensate for oversize and undersize cutter diameters?

Axis Presets

30. Explain the difference between the absolute position register and the command accumulator register.

31. What is the purpose of the zero shift control on the MCU?

32. What is the advantage of an MCU with a full floating zero?

Tool Length Compensation

33. What information should be included on a tool list?

34. How can the actual tool length be entered into the MCU?

35. Why do some industries use preset tooling?

36. What is the purpose of tool assembly drawings?

37. Briefly explain how semiautomatic tool compensation operates.

CHAPTER

SEVEN

Angular and Contour Programming

Linear interpolation involves moving the cutting tool from one position to another in a straight line. With this type of programming, any straight-line section can be machined, including all tapers or angular surfaces. When linear moves are programmed, the coordinates (XY axes) for the beginning and end of each line must be given. Linear interpolation may also be used to produce arcs and circles, but it is not often used for this purpose because of the large number of coordinate locations which must be programmed to keep each segment move as short as possible. The smaller each segment is, the smoother the arc or circle.

The development of circular interpolation has greatly simplified the programming of arcs and circles. The only information required to program an arc up to 90° is the end point of the arc and the arc radius. When programming a complete circle, the radius and four or five coordinate positions must be given.

1. Use trigonometry to calculate the coordinate locations of angular surfaces

2. Program tapers and angles on various workpieces

3. Program arcs and circular forms as required on a variety of parts

ANGULAR PROGRAMMING

Whenever movement is required along two axes (X and Y), the axes move simultaneously along a vector path (Fig. 7-1). The rate of travel along the vector path is set automatically by the machine control unit (MCU) so that it is equal to the programmed feed rate. Since an angle is a straight line connecting a start point and an end point, linear interpolation can be used to produce

Fig. 7-1   Angular surfaces can be programmed by specifying a start point, an end point, and a vector feed rate.

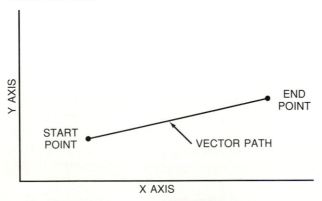

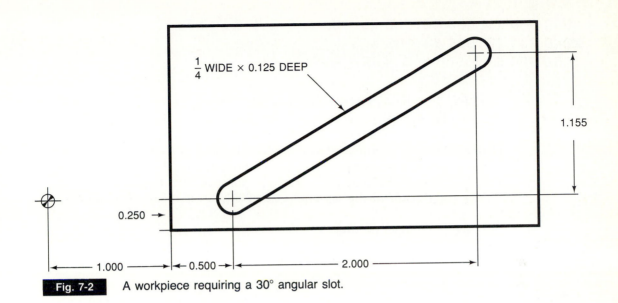

Fig. 7-2 A workpiece requiring a 30° angular slot.

angles and tapers regardless of length. Once the two points have been programmed, along with a vector feed rate, the MCU holds this information in its memory until the end point of the line is reached. Therefore the function of interpolation is to store information and constantly compare and correct the machine axes movement to keep a straight-line movement between the start point and end point coordinates at a specified vector feed rate.

The sample workpiece shown in Fig. 7-2 requires that a 30° angular slot 0.250 in. wide by 0.125 in. deep be cut. Before this operation is programmed, it is first necessary to calculate the coordinate positions (XY) of the start point and end point of the angular slot. If this will be programmed in the incremental mode, the calculations are as follows:

> Start point   X1.500 Y0.250
> End point   X2.000 Y1.155
> Feed rate   10 in./min

The program to machine the workpiece would be as follows:

```
N010 G91
```
Incremental mode.

```
N020 G70
```
Inch programming.

```
N030 G00 X1.500 Y0.250 R0.100 Z−0.125 F 10 M03
N040 G01 X2.000 Y1.155 M06
N050 G00 X−3.500 Y−1.405 M30
```

In Fig. 7-2, the coordinate dimensions of the start point and end point of the angular surface were given on the part print. In Fig. 7-3, a five-sided figure (a pentagon) is required to be machined, and the part print does not clearly define the coordinate locations of each start point and end point; therefore it will be necessary to use simple trigonometry to calculate the XY coordinate location for the start point and end point of each angular surface.

The coordinates of the various points of the pentagon are as follows:

Point 1   X0.000 Y−1.00

Fig. 7-3   A workpiece which requires a five-sided figure (a pentagon) to be machined.

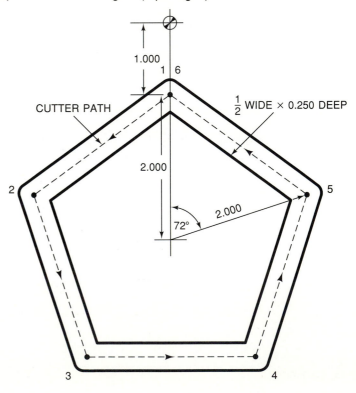

POINT 1 X   0.000   Y −1.000
      2 X −1.902   Y −1.382
      3 X   0.7265 Y −2.236
      4 X   2.351
      5 X   0.7265 Y   2.236
      6 X −1.902   Y   1.382

Point 2   The length of the side opposite and the side adjacent in triangle A must be calculated to find the X and Y coordinate location

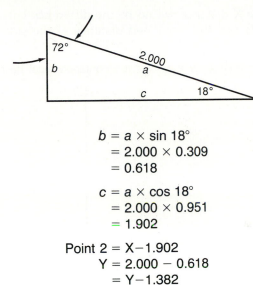

$$b = a \times \sin 18°$$
$$= 2.000 \times 0.309$$
$$= 0.618$$

$$c = a \times \cos 18°$$
$$= 2.000 \times 0.951$$
$$= 1.902$$

$$\text{Point 2} = X{-}1.902$$
$$Y = 2.000 - 0.618$$
$$= Y{-}1.382$$

Since each point of the pentagon is 72° apart, refer to Fig. 7-4 for details of triangle B.

Point #3 =

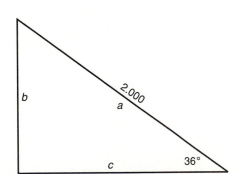

$$b = a \times \sin 36°$$
$$= 2.000 \times 0.5877$$
$$= 1.1755$$

$$c = a \times \cos 36°$$
$$= 2.000 \times 0.809$$
$$= 1.618$$

The X distance would be the difference between the point 2 X dimension (1.902) and the calculated distance of the side opposite (1.1755), or 0.7265.

Fig. 7-4   The coordinate locations of the pentagon must be calculated by using trigonometry.

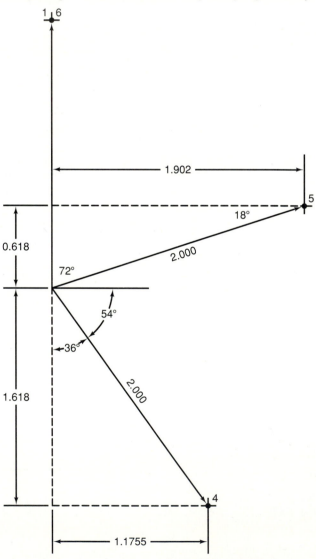

Point 3 = X0.7265

The Y distance would be the calculated length of the side adjacent (1.618) plus the distance point 2 is above the center of the pentagon (0.618), or 2.236.

Point 3   Y−2.236

Point 4   The X axis is the same distance from the center line as point 3; therefore the X location is 1.1755 + 1.1755 or
2.351 = X2.351
The Y axis of points 3 and 4 are the same; therefore there is no Y movement for point 4

Point 5   The X and Y locations for point 5 are the same as for point 4 but on the opposite side of the center line X0.7265.
Y2.236

Point 6   The distance from point 5 to 1 is the same as the distance from point 1 to 2
X−1.902
Y1.382

Programming the Pentagon

Let us assume that a five-sided (pentagonal) groove ½-in. wide and 0.250 in. deep must be machined on the part shown in Fig. 7-3. A two- or three-flute end mill should be used because it is center-cutting and can be used to plunge-cut to depth. The following incremental program is required to machine the part to the required specifications:

```
N010 G91
```
Incremental.

```
N020 G70
```
Inch.

```
N030 G00 X0.000 Y-1.000 R0.100 M03
N040 G01 Z-0.350 F10
N050 X-1.902 Y-1.382
N060 X0.7265 Y2.236
N070 X2.351
N080 X0.7265 Y2.236
N090 X-1.902 Y-1.382 M06
N100 G00 X-0.000 Y1.000 M30
```

CUTTER OFFSET CALCULATIONS

There are two common methods of programming a cutter path: by guiding the edge of the cutter or by guiding the center of the cutter. Generally the center of the cutter is programmed, especially when angular surfaces which are tangent to another surface or a radius are cut. When any surface of a workpiece parallel to the X or Y axis of the machine tool slides is cut, the amount of offset required is the *radius of the milling cutter*. Cutter positions #1 and #2 in Fig. 7-5 show two surfaces where the amount offset is equal to the radius of the cutter.

Fig. 7-5   The cutter center point must be offset to cut various surfaces on a workpiece.

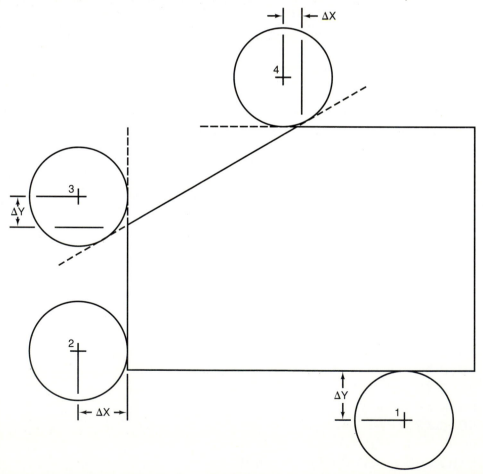

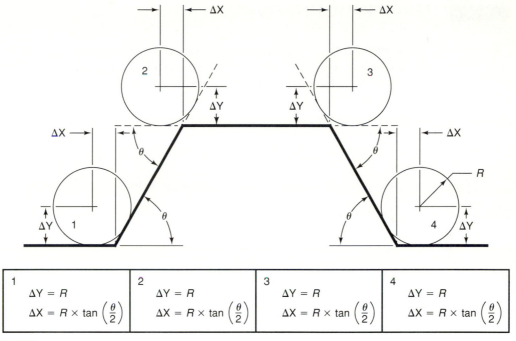

$$\Delta X \qquad\qquad \Delta X$$

1	2	3	4
$\Delta Y = R$ $\Delta X = R \times \tan\left(\dfrac{\theta}{2}\right)$	$\Delta Y = R$ $\Delta X = R \times \tan\left(\dfrac{\theta}{2}\right)$	$\Delta Y = R$ $\Delta X = R \times \tan\left(\dfrac{\theta}{2}\right)$	$\Delta Y = R$ $\Delta X = R \times \tan\left(\dfrac{\theta}{2}\right)$

Fig. 7-6   Commonly used formulas for calculating cutter offset.

When angles on a workpiece are milled, it is necessary to position the cutter in relation to the surface to be cut in order to correctly cut the required angle. The cutter circumference will cut the angle in a different position than when it mills a straight side, so it is necessary to calculate the cutter center point in relation to where the circumference contacts the work.

The coordinate position where an angular surface meets another surface is not the same coordinate position required for the center of the milling cutter. Position #3 in Fig. 7-5 shows that the cutter must be offset a certain distance along the Y axis in order to cut the angular surface. Position #4 shows that when a milling cutter completes an angular cut and must cut a parallel surface, the cutter must again be offset a certain distance along the X axis in order to cut the next surface to the proper size. Notice the difference between coordinate locations of the start and end of the angular surface and the cutter center coordinate locations.

Before any angular surface is programmed, it is necessary to calculate the required offset along the X or Y axis to find the amount that the cutter must be offset to produce the required angular surface. Figure 7-6 shows a variety of formulas which can be used to calculate the cutter offsets along the X or Y axis for angular surfaces.

PROGRAMMING THE TRIANGLE

The triangle in Fig. 7-7 must be machined to the dimensions shown. A 3-in.-diameter shell milling cutter will be used so that the entire surface of the part can be machined in one pass. Before this operation is programmed, it is necessary to calculate the cutter offsets for the four positions of the cutter shown in Fig. 7-8.

Fig. 7-7   The dimensions of the triangle which must be cut in the milling and drilling exercise.

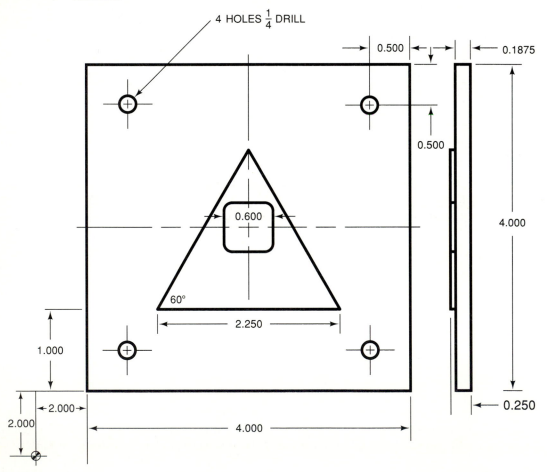

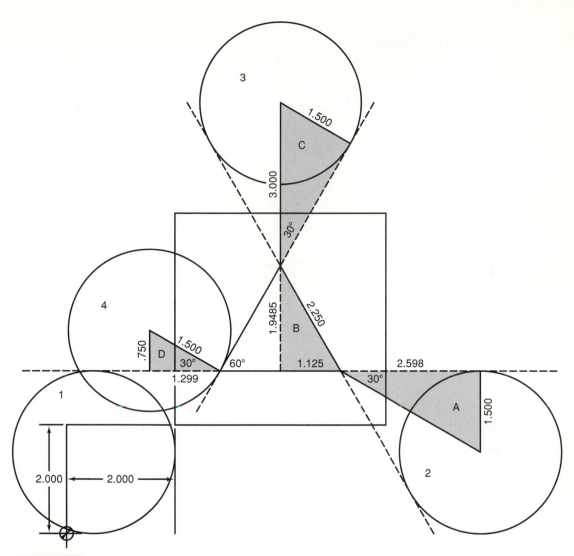

Fig. 7-8 The cutter offsets for the triangle in stage 3 of the milling and drilling exercise must be calculated.

Position #1

The cutter location to mill the lower edge of the triangle is calculated as follows:

- The XY zero is 2.000 in. below and to the left of the workpiece
- The cutter diameter is 3.000 in. (the radius is 1.500 in.)
- The lower edge of the triangle is 1.000 in. from the lower edge of the workpiece

$$X = 2.000 - 1.500$$
$$= 0.500$$

$$Y = (2.000 - 1.500) + 1.000$$
$$= 1.500$$

Position #2

The cutter location for the right side of the triangle is calculated as follows:

- The X axis offset must be calculated for triangle A:

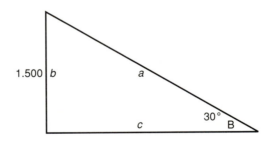

$$c = b \times \cot B$$
$$= 1.500 \times 1.732$$
$$= 2.598$$

$$X = 1.500 + 0.875 + 2.250 + 2.598$$
$$= 7.223$$

$$Y = 0.0$$

Position #3

The cutter location for the left side of the triangle is calculated as follows:

- The height of triangle B must be calculated:

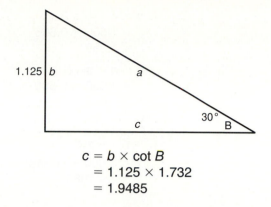

$$c = b \times \cot B$$
$$= 1.125 \times 1.732$$
$$= 1.9485$$

- The Y axis offset must be calculated for triangle C:

$$a = \frac{b}{\sin B}$$

$$= \frac{1.500}{0.5}$$

$$= 3.000$$

$$X = 2.598 + 1.125$$
$$= -3.723$$

$$Y = 1.500 + 1.9485 + 3.000$$
$$= 6.4485$$

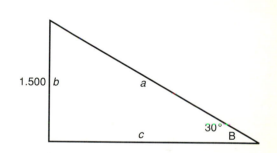

Position #4

The cutter location for the end of the left side of the triangle is calculated as follows:

- The X and Y axis offset must be calculated for triangle D:

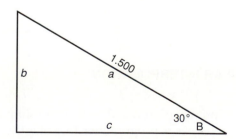

$$c = a \times \cos B$$
$$= 1.500 \times 0.866$$
$$= 1.299$$

$$b = a \times \sin B$$
$$= 1.500 \times 0.500$$
$$= 0.750$$

$$X = 1.125 + 1.299$$
$$= -2.424$$

$$Y = 3.000 + (1.9485 - 0.750)$$
$$= -4.1985$$

Position #5

The XY zero, or home, is calculated as follows:

$$X = 2.000 - (1.299 - 0.875)$$
$$= -1.576$$

$$Y = 0.750 + 1.000 + 2.000$$
$$= -3.750$$

The Program for the Triangle

This program for machining the triangle on the NC machining exercise follows the program for milling the recess in Chapter 6. The material of the workpiece is aluminum, and the programming mode is still incremental.

```
N250 G00 X0.500 Y1.500 R0.100 Z-0.1625 S400 F10 M03
N260 G01 X7.223
N270 X-3.723 Y6.4485
N280 X-2.424 Y-4.1985
N290 G00 X-1.576 Y-3.750 M06
```

CIRCULAR INTERPOLATION

Circular interpolation was developed to simplify the programming of arcs and circles. It allows a programmer to make the cutting tool follow any circular path ranging from a small arc segment to a full 360° circle. All that has to be

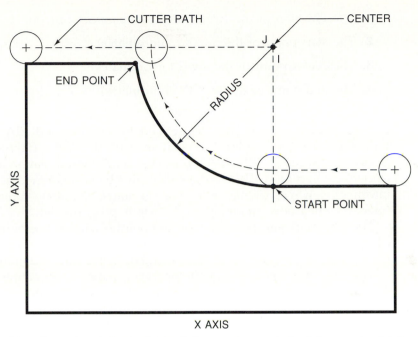

Fig. 7-9 Two-dimensional circular interpolation requires a start point, an end point, the coordinate location of the arc center, and the direction of cutter travel to be programmed.

programmed for an arc or circle on some MCUs are the coordinate locations of the start point and end point of the arc, the radius of the circle, the coordinate location of the circle center, and the direction in which the cutter is to travel (Fig. 7-9). The circular interpolation of the MCU automatically breaks up the arc into very small linear moves, generally 0.0001 or 0.0002 in. (0.0025 or 0.005 mm) each, to describe the circular path. The MCU then generates the controlling signals to move the cutting tool to produce the desired arc or circle. If the same arc or circle were to be programmed in linear interpolation, hundreds or even thousands of coordinates, each defining a span, would have to be programmed.

There are various control units on the market; some can generate only one quadrant at a time, while others can generate a full circle. Some models of MCUs are limited to circular interpolation in a two-axis plane at a time, such as XY, XZ, or YZ axes, while others can interpolate circular movements for three axes at the same time. Circular interpolation can also be used to generate second- and third-degree curves and free-form shapes which can be closely described with a series of arcs or circles.

When circular interpolation is programmed, four pieces of information are necessary:

1. The direction of the cutter travel (preparatory function)

2. The start point of the arc (XY coordinates)

3. The center point of the arc (I J coordinates)

4. The end point of the arc (XY coordinates)

The *direction of cutter travel* is defined by the standard EIA preparatory function codes for circular interpolation. G02 is circular interpolation in a clockwise direction (CW). G03 represents circular interpolation in a counterclockwise direction (CCW). These codes must be programmed in the block of information where circular interpolation starts, and they remain effective (modal) until a new preparatory (G) code is programmed.

The *start point of arc* is usually the end point of a linear line or the end point of a previous arc. The start point positions the cutting tool to start machining the arc and is generally given as XY and/or Z coordinate dimensions.

The *center point of the arc* (XY and/or Z coordinates) is the center of the circle or arc and is described by I (X coordinate value), J (Y coordinate value), and K (Z coordinate value) (Fig. 7-9). Generally the I, J, and K words are incremental values regardless of whether they have been programmed in the absolute or incremental mode. The I and J coordinate values are always taken *from the start point of each arc or 90° segment to the center point of the radius.*

The *end point of the arc* (XY and/or Z coordinates) is the last point where the cutter path center line completes the circular path. Whenever an arc uses more than one 90° quadrant, the point where it crosses into the next quadrant must be programmed as the end point. The MCU assumes that this is also the start point for the next quadrant; therefore it is only necessary to program the end point for the next quadrant.

Programming an Arc

The two most common methods of programming an arc are by center point programming and by radius programming. Some MCUs will generate an arc if it is defined by the arc center point (XY coordinates) and the end point of the arc. Other MCUs will generate an arc if it is defined by the arc radius and the end point of the arc. Before an arc or circle is programmed, it is necessary to determine what information is needed for the particular MCU being used.

There are three questions which must be answered before an arc is programmed. Place a pencil point at the start point of the arc and answer the following questions before the pencil point is moved.

1. *What way?*   Clockwise (G02) or counterclockwise (G03) direction from the start point of the arc.

2. *Where to?*   The X and Y coordinates of the end point of the arc.

3. *How far?*   The I and J values from the start point of the arc to the center of the circle.

The information required for center point programming is as follows:

1. *G-code*   G02 for circular interpolation clockwise, G03 for circular interpolation counterclockwise.

2. *End point*   The X and Y coordinates of the end point of the arc.

3. *Center point*   The coordinates of the center point of the arc. The letters I (X axis) and J (Y axis) are used to define the point.

The information required for radius programming is as follows:

1. *G-code*   G02 for circular interpolation clockwise, G03 for circular interpolation counterclockwise.

2. *End point*   The X and Y coordinates of the end point of the arc.

3. *Radius*   The radius of the arc preceded by the letter address R.

To Program an Arc

For the two arcs shown in Fig. 7-10 to be programmed, the start point and the end point for each arc must be programmed. If the incremental mode is used, the programming would be as follows:

```
H010 %
N020 G91
N030 G70
N040 G00 X2.500 Y0.500 R0.100 M03
```
Rapid to position #1.
The spindle rapids down to 0.100 above the work surface.

```
N050 Z-0.350 F2.0
```
The cutter feeds into the workpiece 0.250 (0.350 − 0.100) at a 2-in. feed rate.

```
N060 G02 X-1.000 Y1.000 I0.0 J1.000 F10.0
```
G02 Circular interpolation clockwise.

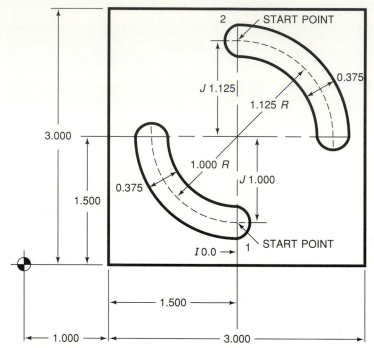

Fig. 7-10 The I and J locations are taken from the start point of the arc to the center of the radius.

X−1.000
Y1.000 } Locates the coordinate position of the end point of the arc.

I0.0
J1.000 } The center of the radius is located from the start point of the arc.

N070 G00 R0.100

The cutter rapids up to 0.100 above the work surface.

N080 X1.000 Y1.125

Rapid to position #2.

N090 G01 Z−0.350 F2.0

The cutter feeds to depth.

N100 G02 X1.125 Y−1.125 I0.0 J−1.125 F10

Same as sequence #N060.

```
N110 G00 R0.100
```
>   Same as sequence #N070.

```
N120 X-3.625 Y-1.500 M06
```
>   Rapid back to XY zero (the start position).
>   M06 stops the spindle and retracts it to the full retract position.

```
N130 M30
```
>   Rewinds the tape in preparation for the next part.

```
N140 %
```
>   End of program.

To Program a Circle

A full circle consists of four 90° arcs, and since many MCUs generate only one quadrant at a time, each arc must be programmed. Since the end point of one arc automatically becomes the start point of the next arc, a full circle requires the *first start point* and *four end points* to be programmed.

The program for the circular groove shown in Fig. 7-11 is as follows:

```
H010 %
N020 G91
N030 G70
N040 G00 X1.000 R0.100 M03
```
>   Rapid to position #1.

```
N050 G01 Z-0.350 F2.0
N060 G02 X2.000 Y2.000 I2.000 F10.0
```
>   Circular groove cut from point 1 to 2.

```
N070 X2.000 Y-2.000 J-2.000
```
>   Circular groove cut from point 2 to 3.

```
N080 X-2.000 Y-2.000 I-2.000
```
>   Circular groove cut from point 3 to 4.

```
N090 X-2.000 Y2.000 J2.000
```
>   Circular groove cut from point 4 to 5.

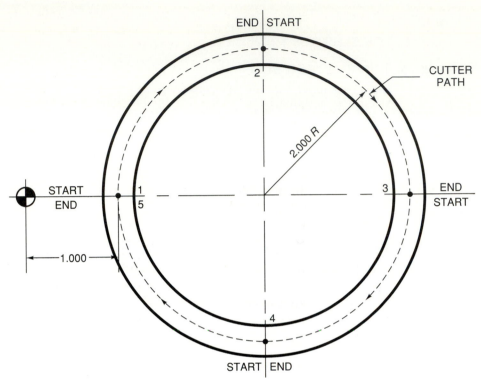

Fig. 7-11 For a full circle to be machined, the *start point* and the end point of each quadrant (90°) must be programmed.

```
N100 G00 R0.100
N110 X-1.000 M06
N120 M30
N130 %
```

SUBROUTINES AND MACROS

A parametric *subroutine*, sometimes called a "program within a program," is used to store frequently used data sequences (one block or a number of blocks of information), which can be recalled from memory as often as required by a Code or call statement in the main program (Fig. 7-12). Subroutines are usually stored separate from, but in the same general area as, a part program. An example of a subroutine could be a drilling cycle where a series of ⅜-in.-

(9.5-mm-) diameter holes 1-in. (25-mm) deep must be drilled in a number of locations on a workpiece. The G81 to G89 series of preparatory functions set up a machining center for various types of automatic machining functions which are called *fixed* or *canned cycles*. The most common fixed cycles are described fully in Chapter 5.

A macro or subroutine is a miniprogram within a main program, which would have to be repeated a number of times on a workpiece. It is a group of instructions or data which is permanently stored in memory and can be recalled as a group to *solve recurring problems* such as bolt-circle locations, drilling and tapping cycles, and other frequently used routines. The macro must be programmed in the incremental mode regardless of whether the main program is in the absolute or incremental mode. The macro must be written as a complete program and stored as P1, P2, etc., and recalled into the main program as required by the code Q1, Q2, etc. An example of a macro would be the XY locations of the three slots in the workpiece illustrated in Fig. 7-12.

When the diameter of a bolt circle and the number of holes on the circle are provided, the MCU can make all the calculations for hole locations and cause the machine tool slides to move into the proper position for each hole. MCUs with this feature can save up to 50 percent in programming time and one-third of the data processing time, and reduce the length of tape required for the program.

Programming a Macro or Subroutine

When the macro (three slots) in Fig. 7-12 is programmed, the following points should be kept in mind:

Fig. 7-12   Macro programming the three slots saves time and computer memory space.

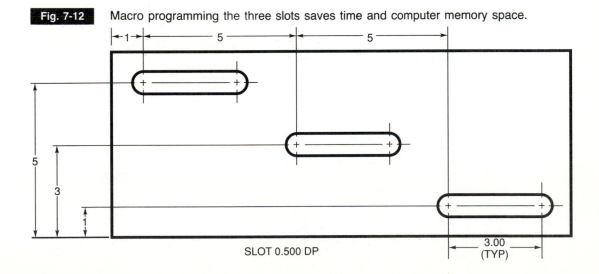

SLOT 0.500 DP

1. The macro must be programmed in the incremental mode (G91).

2. Temporarily stored subroutines (macros) are usually included at the beginning of the main program.

3. The subroutine or macro program is usually stored under a call statement such as PGM1 or P1 (program #1) and recalled into the main program by a CLS1 or Q1 call statement.

4. The sequence numbers of the macro should be large enough so that they do not conflict with the sequence numbers of the main program.

5. If the absolute programming mode (G90) is used in the main program, the G90 code must be used at the end of the macro in order to return to the programming mode of the main program.

6. The M02 code should be used at the end of the macro to reset the macro so that it can be used again as required.

The Macro for the Three Slots (Stored as P1)

```
N1000 P1
```

```
N1010        (Macro #1–3 slots)
```

```
N1020 G91
```
incremental

```
N1030 G01 Z-0.600 F1.0
```
G01—linear interpolation.
Z-0.600—the cutter is fed down into the work 0.500 + 0.100 gage height.
F1.0—the feed rate is 1 in./min.

```
N1040 X3.000 F3.0
```
X3.000—a slot is cut 3.000 in. long along the X axis.
F3.0—the feed rate for milling the slot is 3 in./min.

```
N1050 G00 Z0.600
```
The cutter rapids out of the workpiece 0.600 in. or back to gage height (0.100 above the work surface).

```
N1060 G90
```
Absolute to return to main program.

N1070 M02

> Rewinds the subroutine.

The Main Program for the Three Slots

N010 %
N020 G90

> absolute

N030 G70
N040 G92 X10.0 Y8.0 T01 H01 M03

> G92—rapids the table from home position to the program XY zero.
> T01—the number and type of cutting tool used.
> H01—the tool length offset for the cutting tool.

N050 G00 X1.0 Y5.000 Z0.100 M08

> G00—the table rapids to position #1.
> Z0.100—the spindle rapids to within 0.100 of the work surface.
> M08—the coolant is turned on.

N060 Q1

> The macro is recalled to mill slot #1.

N070 X6.000 Y3.000

> The table rapids to position #2.

N080 Q1

> The macro mills slot #2.

N090 X11.000 Y1.0

> The table rapids to position #3.

N100 Q1

> The macro mills slot #3.

N110 X10.000 Y8.000 Z2.000 M05

> The table rapids back to the program XY zero.
> Z2.000—the spindle rises 2.000 in. above the work surface.
> M05—stops the spindle.

N120 M30

> Rewinds the tape in preparation for the next part.

To Program the Four Circular Grooves of the NC Machining Exercise

The four circular grooves on the NC machining exercise are all the same diameter, width, and depth, and therefore they should be entered as a subroutine or macro to save programming time and computer memory. Once this macro program (entered as a separate program from the main program) is completed, it is only necessary to position the machine table (XY coordinates) at the start point of each circular groove and recall the macro into the main program to cut the groove.

Fig. 7-13   The four circular slots should be programmed as a subroutine or macro.

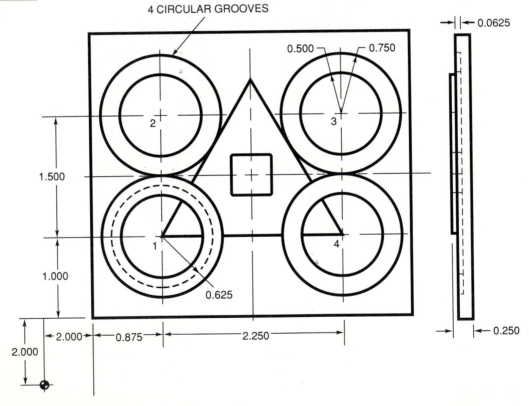

To Program the Circular Grooves

The program for machining the four circular grooves in Fig. 7-13 follows the milling of the triangle covered earlier in this chapter.

Note:

1. A ¼-in.-diameter two-flute end mill should be used to machine the circular grooves.

2. The length of the end mill must be preset to the reference plane 0.100 in. above the work surface.

3. The start point of each groove will be on the left-hand side of the center line.

The Macro Program (Stored as P1)

```
N1000 P1
N1010
```
(Macro #1—Circular Grooves)

```
N1020 G91
```
Incremental

```
N1030 G01 Z-0.225 S3000 F1.0
```
G01—linear interpolation.
Z−0.225—the cutter is fed down into the work 0.125 + 0.100 gage height.

```
N1040 G02 X0.625 Y0.625 I0.625 J0.0
N1050 X0.625 Y-0.625 I0.0 J-0.625
N1060 X-0.625 Y-0.625 I-0.625 J0.0
N1070 X-0.625 Y0.625 I0.0 J0.625
```
N1040 to N1070 The circular groove is cut one segment at a time.

```
N1080 G00 Z0.100
```
The cutter rapids out of the work to 0.100 above the surface.

```
N1090 M02
```
The macro is rewound in preparation for cutting the next groove.

The Main Program for Milling the Circular Grooves

N300 G00 X2.250 Y3.00

> The table rapids to the start point of circular groove #1.

N310 Q1

> The macro is recalled into the main program and the first circular groove is cut.

N320 G00 Y1.500

> The table rapids to the start point of circular groove #2.

N330 Q1

> The second circular groove is cut.

N340 G00 X2.250

> The table rapids to the start point of circular groove #3.

N350 Q1

> The third groove is cut.

N360 G00 Y−1.500

> The table rapids to the start point of circular groove #4.

N370 Q1

> The fourth groove is cut.

N380 G00 X−4.500 Y−3.000 M06

> The table rapids back to the XY zero.
> M06—the spindle is turned off and rises to the retract position.

Programming for the Holes

Circular grooves #1 and #2 require seven equally spaced ⅛-in.-diameter holes to be drilled on a 1.250-in.-diameter bolt circle. Circular grooves #3 and #4 require nine equally spaced ⅛-in.-diameter holes to be drilled on a 1.250-in.-diameter bolt circle. Both the seven and the nine holes in Fig. 7-14 require a separate macro, and the coordinate locations (XY) for every hole must be calculated by trigonometry or by using the Woodworth coordinate tables.

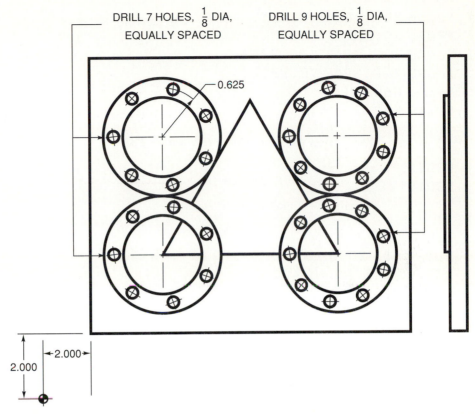

DRILL 7 HOLES, $\frac{1}{8}$ DIA, EQUALLY SPACED

DRILL 9 HOLES, $\frac{1}{8}$ DIA, EQUALLY SPACED

0.625

2.000

2.000

Fig. 7-14 Two subroutines (macros) are required: one for the seven-hole circle and the other for the nine-hole circle.

The Macro Program for Seven Holes (Stored as P2)

```
N2000 P2
N2010
        (Macro #2—7 Holes)

N2020 G91
N2030 G81 X0.0 Z-0.370 R0.100 S6000 F1.0
N2040 X0.2353 Y0.4886
N2050 X0.5287 Y0.1207
```

```
N2060 X0.4241 Y-0.3382
N2070 X0.0 Y-0.5423
N2080 X-0.4241 Y-0.3382
N2090 X-0.5287 Y0.1207
N2100 G80 M06
N2110 M02
```

The Macro for Nine Holes (Stored as P3)

```
N3000 P3
N3010
```

(Macro #3—9 Holes)

```
N3020 G91
N3030 G81 X0.0 Y0.0 Z-0.370 R0.100 S6000 F1.0
N3040 X0.1462 Y0.4017
N3050 X0.3702 Y0.2138
N3060 X0.4211 Y-0.0743
N3070 X0.2748 Y-0.3275
N3080 X0.0 Y-0.4275
N3090 X-0.2748 Y-0.3275
N3100 X-0.4211 Y-0.0743
N3110 X-0.3702 Y0.2138
N3120 G80 M06
N3130 M02
```

The Main Program for Drilling the Holes

The program for drilling the seven and nine holes in Fig. 7-14 follows the milling of the circular grooves covered earlier in this chapter.

Note:

1. A short stubby ⅛-in.-diameter drill should be used for drilling the holes.

2. The length of the drill must be preset to the reference plane 0.100 in. above the work surface.

3. The start hole for each bolt circle is the left-hand hole on the center line.

4. The holes will be drilled in a clockwise direction.

The Program

N390 G00 X2.250 Y3.000

The table rapids to the start hole in circular groove #1.

N400 Q2

The seven-hole macro is recalled and the holes are drilled.

N410 G00 X-0.2353 Y1.9887

The table rapids from hole #7 of circular groove #1 to the start hole in circular groove #2.

N420 Q2

The seven-hole macro is recalled and the holes are drilled.

N430 G00 X2.0147 Y0.4887

The table rapids from hole #7 of circular groove #2 to the start hole in circular groove #3.

N440 Q3

The nine-hole macro is recalled and the holes are drilled.

N450 G00 X-0.1462 Y-1.0982

The table rapids from hole #9 of circular groove #3 to the start hole in circular groove #4.

N460 Q3

The nine-hole macro is recalled and the holes are drilled.

N470 M30

End of tape command, rewinding the tape in preparation for the next part.

REVIEW QUESTIONS

1. What type of forms or surfaces are generally produced by linear interpolation?

2. For what purpose is circular interpolation used?

Angular Programming

3. Define the term *vector path*.

4. What three factors are required in order to program an angle?

5. Calculate the coordinate locations for the three sides of the triangle.

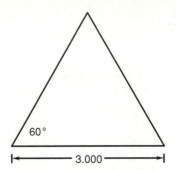

6. Why should a two- or three-fluted end mill be used when milling grooves or slots?

Cutter Offset Calculations

7. Name two common methods of programming a cutter path.

8. How much offset is required when surfaces parallel to the X or Y axis must be machined?

9. Why is it necessary to offset a cutter when milling an angular surface?

10. Calculate the cutter offsets required for a 1-in.-diameter end mill at points #1 and #2 to cut the angular surface.

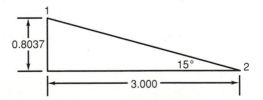

Circular Interpolation

11. Why was circular interpolation developed?

12. How does the MCU break up an arc for circular interpolation?

13. Name the difference between the various control units for circular inter-polation.

14. What four pieces of information are necessary when programming for circular interpolation?

15. Define G02 and G03.

16. Explain the relationship of I and J to the XY axis.

17. Name two methods of programming an arc.

18. What three questions should be answered before programming an arc?

19. List the information required for:
   (a) Center point programming
   (b) Radius programming

20. On machines which can only generate a 90° arc or quadrant at a time, what information is necessary in order to program a complete circle?

Subroutines and Macros

21. Define a subroutine.

22. Why are subroutines or macros useful?

23. In what mode must subroutines be programmed?

24. How are macros written, stored, and recalled into the main program?

25. Explain the importance of the following in a subroutine.
   (a) Large sequence numbers
   (b) G90 code
   (c) M02 code

26. How can the XY coordinates of holes on a bolt circle be calculated?

CHAPTER

EIGHT

Miscellaneous Numerical Control Functions

Since the early days of numerical control (NC), there has been a steady refinement of the basic process to a point at which it is an invaluable and indispensable manufacturing tool. Advanced computer technology, new machine control unit (MCU) features, and the development of sophisticated NC software programs are having a major impact on part programming. Virtually any form can now be generated by NC because of numerous developments, among them, circular, helical, and parabolic interpolation; contouring; digitizing; and scanning.

After completing
this chapter,
you should be
able to:

1. Recognize the most common NC programming languages

2. Understand the function of general and postprocessors

3. Know the purpose of digitizing, scanning, and graphics in NC

4. Identify and select NC tooling

NC PROGRAMMING LANGUAGES

There is a large variety of computer languages (close to a hundred) available for NC programming. However, many of them are very specialized and, as a result, not widely used. It is advisable to learn a language that is widely used because:

1. A more popular language can be programmed on many types of computers.

2. More information and more software programs are available.

3. A popular program eases communication with many computers.

4. Training can be obtained from many sources.

5. The computer language is more likely to be around for a long time.

The seven most common NC programming languages are APT, ADAPT, UNIAPT, NUFORM, SPLIT, ACTION, and COMPACT II (Fig. 8-1). Each will be briefly discussed.

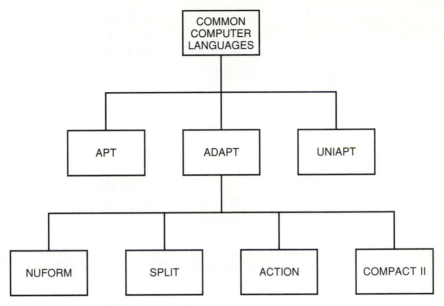

Fig. 8-1 Common types of NC programming languages.

APT

Automatic Programming of Tools (APT) is the oldest and one of the most powerful NC processor languages, initially developed by a group at MIT's Electronic Systems Laboratory in 1956–1957. The scope of APT was later enlarged by the IIT Research Institute Manufacturing Productivity Center in Chicago in conjunction with a group of APT sponsors and manufacturers. APT is generally used on large-capability computers and can perform the mathematics required for complex curves using four- or five-axis contouring techniques.

Characteristics of APT

1. An English-type language is used to describe the geometric points and surfaces of a part on an engineering drawing.
   - Up to 17 geometric facts, such as points, circles, cones, and surfaces, may be defined, each in several forms of description.
   - The words are limited to six letters, each with a specific and precise meaning, such as:

ATANGL	at angle to
C_1	circle #1
FEDRAT	feed rate
GOFWD	go forward
GOLFT	go left
GORGT	go right
GOTO	go to
L_1	line #1
PL	plane
PT_1	point #1
SETPT	set point
TANTO	tangent to

2. The cutter path can be described:
   - Point-to-point programming must specify positions in absolute or incremental movements.
   - Continuous path programming guides a specific cutter along a path in a series of motions, each defined by the intersection of two surfaces—the *part surface* (usually the depth of cut) and a *drive surface* (usually the path through space).
   - All cutter path sequences start with an initial position, and continuous path sequences require a tolerance specification.

3. Machine tool functions such as speeds, feeds, coolants, optional stops, and tape rewind can be described.

The words of the APT processor language are precise, explicit, and unchanging. If used properly, they will give specific and highly predictable results.

ADAPT

The ADAPT language was developed as an Adaptation of the APT language. It uses only about one-half of the words used by APT and was designed to run on smaller computers. Its development was sponsored by the U.S. Air Force, and it is used for programming two simultaneous contouring motions in a plane and a third axis of linear control.

UNIAPT

The UNIAPT language is almost the same as the APT language except that there are a few words in each language that are not recognizable or compatible with the others. The UNIAPT language and its processor operate in a batch mode on a local dedicated minicomputer.

UNIAPT can handle programming for all three-axis and most four- and five-axis machine tools. In some cases, an optional multiaxis module is available which allows the UNIAPT system to process continuous path five-axis parts.

NUFORM

The NUFORM language uses Numeric rather than mnemonic (memory) codes; no words, letter codes, abbreviations, or punctuation marks are used. The programmer must place each numeric entry (code numbers or dimension numbers) on an 80-column manuscript in the appropriate column (in 1 of 10 fields). Each manuscript line must then be punched into an 80-column card, and the entire deck of cards is then fed into the computer. The computer will then generate the program as punched cards, punched tape, printed listing, or any combination of these.

The NUFORM language and its processor generally operate in a batch mode on a local dedicated minicomputer. This language differs greatly from other computer language, and great care must be used in making comparisons.

SPLIT

The SPLIT language (Sundstrand Processing Language, Internationally Translated) and its machine-dependent processor operate in either batch or interactive mode on a local computer which may be time-shared or fully dedicated. SPLIT is the parent language of a group which consists of SPLIT, ACTION, and COMPACT II. Though these languages are very similar, the processor for each is very different.

The SPLIT language program has these characteristics:

1. It consists of statements describing the actual operation to be performed, the dimensions necessary for each operation, and the required auxiliary functions.

2. Each statement consists of one or more fields (''major operations'' or ''minor operations'') separated by commas.

3. Each field contains an alphabetic symbol representing its function and a numeric part.

4. Each statement can contain *only one major* operation and none, one, or more minor operations.

ACTION

The ACTION language is very similar to the SPLIT language but uses *very* different processors. ACTION was designed to be easily learned, to reduce part programming time, and still to be able to handle the majority of parts requiring NC programming. The ACTION language and its machine-dependent processor operate in either batch or interactive modes on a local or remote computer. ACTION in a local interactive mode can be upgraded and extended to a direct NC system.

COMPACT II

The COMPACT II language and its machine-dependent processor operate in an interactive environment on remote, time-shared computers. This language and its processor were developed to satisfy the needs of a large majority of programming requirements and also for parts with modest complexity.

COMPACT II is an easy language to learn, and there are only a few points which should be kept in mind:

1. The symbols and language are simple.

2. The words used are easily recognized.

3. The words may be placed in any sequence within a program statement.

4. The statement structure follows accepted NC machining steps or procedures.

5. COMPACT II specifies the machine tool, defines the part shape, selects the tools required, and directs the cutting action.

COMPACT II is widely used throughout industry for milling, turning, drilling, boring, EDM, punch press work, etc.

APT GENERAL PROCESSOR

A *processor* is a set of computer instructions which transforms tool centerline data into machine motion commands using the proper code and format required by a specific MCU. The processor is used to describe the geometry, dynamics, and features of a machine tool to the computer. Included with this may be information on feed rate, speed rate, and auxiliary function commands.

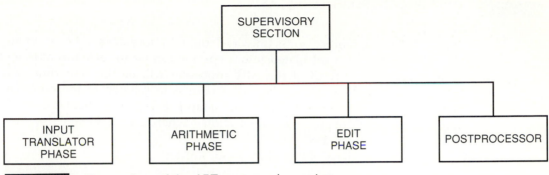

The sections of the APT programming system.

The APT system consists of four sections or phases plus a supervisory section which controls the flow of information (Fig. 8-2). Each section or phase plays an important role in NC programming.

Phase 1: Input Translator
- The translator reads and checks manuscript statements for errors.
- An error signal alerts the programmer of any errors.
- The translator separates source statements and classifies them by the type of operation.
- All the necessary data is extracted, rearranged, and recorded in computer-usable form.

Phase 2: Arithmetic
- Data is received from the input translator phase.
- All the calculations for a given machining problem are made using a built-in library of tables, symbols, and subroutines.
- The calculations are made to find the coordinate values of the cutting tool's center point.

Phase 3: Edit
- Data is received from the arithmetic phase and has three major purposes:
  a. Vertical tool axis orients the spindle from the vertical position.
  b. TRACUT and COPY transform and manipulate the data from phase 2.

Phase 4: Postprocessor
- Data is received from the edit phase and is converted into the language format for each particular machine tool control unit.

POSTPROCESSORS

The APT language processor is universal and therefore does not convert any data which has been calculated into a specific format for any machine tool control unit. The output of the APT processor will be the centerline (CL) which tells where the CL of the cutter path is in relation to the part. The postprocessor takes this CL output and adapts it to each particular machine tool control unit. The postprocessor section is where the actual tape used on a machine tool is generated (Fig. 8-3).

The main functions of a postprocessor are to:

1. Convert cutter CL data into coordinate dimensions

2. Output preparatory and miscellaneous functions

3. Calculate cutter compensation data

4. Generate circular and parabolic points

5. Control the part size by controlling the amount of overshoot of the machine slides

6. Ensure that the machine tool physical limits such as range or feed rate are not exceeded

Fig. 8-3 The postprocessor adapts the format of APT programs to suit each particular machine tool control unit. (*Cincinnati Milacron, Inc.*)

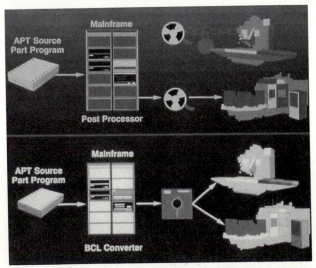

Because of the large number of postprocessors that are available, it is important for the programmer to become familiar with the documentation required in order to produce the desired results.

APT PROGRAMMING

The APT vocabulary consists of about 260 "words" (80 major words and 180 modifiers) including punctuation. The words are limited to six alphanumeric characters, one of which must be alphabetic. The APT words are precise and explicit and do not have shades of meaning as many words in the English language do.

APT programs consist of a series of statements which are listed in an orderly fashion on a manuscript. Statements are used for four purposes:

1. To define geometric points and surfaces on a part which would represent the part's size and shape

2. To describe a cutter and its path—this statement moves the cutter to the various points along the geometric path

3. To describe auxiliary machine tool functions such as spindle rotation, spindle speed, feed rate, coolant, etc.

4. To compute—making all the arithmetic, trigonometric, vectorial, and geometric calculations necessary to produce the part

Statements are made up of a major word followed by a slash (/) and some minor information following the slash. The program for the simple move shown in Fig. 8-4 would be as follows:

1. $GOLFT/HL_1, TANTO, C_1$

Go left (GOLFT) along horizontal line #1 (HL_1) and stop where HL_1 is tangent to (TANTO) circle #1 (C_1)

2. $GOFWD/C_1, TANTO, VL_1$

Continue in a forward direction (GOFWD) along circle #1 (C_1) and stop where C_1 is tangent to (TANTO) vertical line #1 (VL_1)

Continuous Path Motion

All low-level languages require very *explicit* (clear or detailed) instructions for continuous path motion. In APT programming, the instructions can be either

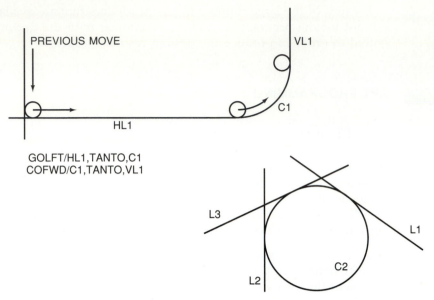

GOLFT/HL1,TANTO,C1
COFWD/C1,TANTO,VL1

C2 = CIRCLE/XLARGE,L2,YSMALL,L3,XSMALL,L1

Fig. 8-4 APT programming statements for simple motion and geometric placement. (*Allen-Bradley*)

explicit or *implicit* (implied or understood). The APT processor looks ahead, and the next *drive surface* becomes the check or stop surface for the current move. Samples of explicit and implicit APT programming are shown in Fig. 8-5.

Example #1 The instructions are very explicit, telling the cutting tool to move along one surface and stop when it contacts another surface.

$$GO/C_1, \ PL_1, \ C_3$$

Locates the cutting tool on plane #1 (PL_1) where circle #1 (C_1) and circle #3 (C_3) intersect

$$TLLFT, \ GOLFT/C_1, \ TO, \ C_2$$

The cutting tool is to travel left (GOLFT) from the start of circle #1 (C_1) to where it intersects with circle #2 (C_2)

$$GOLFT/C_2, \ TO, \ C_3$$

The cutting tool travels left from the start of C_2 to where it intersects C_3

$$\texttt{GOLFT/C}_3, \ \texttt{TO}, \ \texttt{C}_1$$

The cutting tool travels left from the start of C_3 to where it intersects C_1

Example #2   The instructions are implicit because APT looks ahead and sees that the second surface to move along is C_2, and therefore it stops its move on C_1 when the tool gets to C_2, etc.

$$\texttt{GO/C}_1, \ \texttt{PL}_1, \ \texttt{C}_3$$

Locates the cutting tool on PL_1 where C_1 and C_3 intersect

$$\texttt{TLLFT}, \ \texttt{GOLFT/C}_1$$

The cutting tool is to travel left along C_1

Fig. 8-5   APT contour explicit and implicit programming statements. (*Allen-Bradley*)

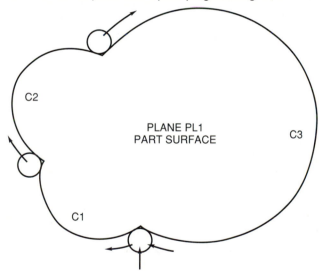

EXAMPLE 1: GO/C1,PL1,C3
         TLLFT,GOLFT/C1,TO,C2
         GOLFT/C2,TO,C3
         GOLFT/C3,TO,C1

EXAMPLE 2: GO/C1,PL1,C3
         TLLFT,GOLFT/C1
         GOLFT/C2
         GOLFT/C3,TO,C1

$$\text{GOLFT/}C_2$$

The cutting tool continues to travel left along C_2

$$\text{GOLFT/}C_3\text{, TO, }C_1$$

The cutting tool continues to travel left along C_3 until it reaches C_1

Example #3 The program can be further shortened by using the implied modal motion feature.

$$\text{GO/}C_1\text{, PL}_1\text{, }C_3$$

Locates the cutting tool on PL_1 where C_1 and C_3 intersect

$$\text{TLLFT, GOLFT/}C_1/C_2/C_3\text{, TO, }C_1$$

The cutting tool will travel left from C_1 through C_2 and C_3 and stop when it comes back to the start point at C_1

ADDITIONAL PROGRAMMING TECHNIQUES

Most part programming consists of a person working from engineering data and developing a manuscript which contains all the information necessary to machine the part. Many other techniques have been developed over the years to make the programming of parts with complex design easier and more accurate. Some of these processes include techniques such as digitizing, scanning, and graphics.

Digitizing

The digitizing process allows an NC part program to be made directly from a scaled engineering drawing of the desired part. Most digitizers consist of a table and a probe which may be connected to one or more arms. The probe in turn is connected to transducers. The scaled engineering drawing is located on the table, and as the probe is passed over a location on the drawing, the coordinate location of each point is recorded. The transducer records the location of each point on the part drawing and upon command transfers this to a tape punch or other processing unit. Many of the digitizers now on the market allow the programmer or operator to enter auxiliary functions, establish reference points, or insert additional information into the program.

Digitizers are especially valuable on parts such as circuit boards, which may require several hundred holes to be drilled in a relatively small space

Fig. 8-6 A digitizer allows an NC program to be made directly from an engineering drawing. (*Hewlett Packard*)

(Fig. 8-6). It would be almost impossible to properly dimension the location of each hole or slot, and therefore drawings for circuit boards or comparable parts are never fully dimensioned. The coordinate locations of each hole or point on the drawing are recorded by placing the reticle (viewing instrument) or probe over the drawing location and then pushing a "record" button. This procedure is repeated at each hole or slot location until the coordinate location of every position has been digitized. The programmer then has the option of including any additional information to the program before the punched tape for the entire program is produced.

Many digitizers have the capability of working with drawings which may be 10 or more times larger than the actual size of the part. When a drawing is larger than the actual part size, any dimensional errors on the part drawing will be reduced by the same factor. For example, an error of 0.010 in. on a drawing which has been enlarged by 10 times would result in an error of only 0.001 in. on the part.

Scanning

Many products such as automobile bodies are designed for aesthetics—an eye appeal which is pleasing to the consumer—and not to basic mathematical

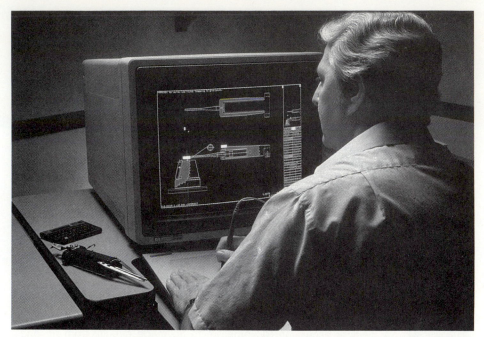

Scanning allows complex contour NC programs to be made by going over the surface of a part with a probe. (*Hewlett Packard*)

formulas or data. Initially, styling is the most important factor in the design, and mathematics becomes involved only after a specific design has been selected.

Taking an approved automobile design and calculating the coordinates necessary to produce an NC program for its reproduction has always been a difficult and tedious task. Scanning (Fig. 8-7) makes it very simple to produce an NC program for very complex forms which would be very difficult, if not impossible, to produce mathematically. A model of the part is made, and the surface of the model is traced with a probe. Analog path data is fed from the probe into a computer which automatically calculates enough digital points (coordinates) to produce a satisfactory NC program. This NC program can then be used to produce the die which can be used to manufacture thousands of similar parts.

On many scanning and digitizing units, it is possible to edit or revise the basic data gathered by the probe or reticle and alter the design of surface. Therefore it is not necessary that the model be made to exact sizes or specifications since it is easy to revise the design somewhat on a scanning or digitizing unit.

Graphics

Computer graphics was first created as a design tool for the product designer. This feature very quickly changed the procedure of designing a product: instead of using conventional drawing instruments and a pencil to draw the part, the designer uses a computer-based system to create a part on a CRT screen. As every change or addition is made by the product designer, the corresponding change or addition appears graphically on the CRT screen. Therefore the designer can see instantly how the final part would look. The designer can also enlarge, rotate, change designs, and complete many other design functions on the part in a fraction of the time it would take by traditional drawing methods.

Programming graphics system (Fig. 8-8) quite naturally evolved from the computer graphics design system to produce NC programs for the part design created on the CRT screen. Once the final part has been developed on the CRT screen, the designer instructs the cutter how to go about machining the part. The computer then produces the NC part program from the geometric dimensions of the part displayed on the screen.

NC TOOLING SYSTEMS

An NC machine tool represents a large investment for any industry, and naturally it is expected that this machine will help to improve productivity and reduce manufacturing costs. For the best operating results to be obtained

Fig. 8-8 On a programming graphics system, an NC program or tape can be made directly from the part design which appears on the CRT screen. (*Bausch & Lomb*)

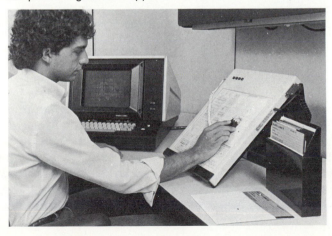

Fig. 8-9 A variety of cutting tools mounted in an indexed turret. (*Kennametal Inc.*)

from an NC machine, it is very important that *proper cutting tools* be used and that the *work be held accurately and securely* in a suitable fixture or workholding device. No machine tool can be very productive without a great deal of thought and planning going into the selection, use, and care of cutting tools. Good cutting tool selection is essential for getting the best results out of any NC machine tool. It is difficult to maintain the accuracy and productivity expected from NC machine tools with poor cutting tools. The programmer must have a good knowledge of cutting tools and their use in order to select the best possible tools to machine a part.

Types of Cutting Tools

There are three major categories of cutting tools commonly used in industry for various NC machine tools:

1. *Single-point tools* for turning and chucking centers, and for boring bar operations

2. *Multipoint fluted tools* for milling, drilling, reaming, countersinking, and counterboring operations on machining centers and on milling and drilling machines.

3. *Special-purpose tools* such as taps, grinding tools, broaches, etc.

Single-Point Tools   Single-point cutting tools are used primarily in turning and chucking centers for a wide variety of machining operations. The cutting tools which are generally preset to specific dimensions are mounted in some form of toolpost or turret which are generally indexible (Fig. 8-9).

Single-point cemented carbide cutting tools are generally used on turning centers because of their ability to maintain a sharp cutting edge at high operating temperatures; as a result, they require less frequent sharpening. The spindle of the machine rotates the workpiece, and the cutting tool held in the turret is brought into contact with the workpiece. A shearing action occurs as the cutting tool touches the work and metal is removed from the diameter in the form of chips.

Single-point cutting tools are available in a variety of shapes, depending on the machining operation required. All cutting tools have certain angles and clearances to make them cut efficiently. The various angles ground on a toolbit are called *basic tool angles* and are often termed *tool geometry*. The angles and shape of a single-point cutting tool are generally defined by a sequence of statements called *tool signature* (Fig. 8-10).

Multipoint Tools   Multipoint fluted cutting tools consist of end mills, twist drills, reamers, shell end mills, milling cutters, etc., all of which have more than one cutting edge. The most common multipoint tools have two or four flutes on smaller-size tools; larger sizes generally have more than four flutes (Fig. 8-11). Many multipoint cutting tools are made of high-speed steel; however, the use of cemented carbide inserts and complete cemented carbide tools is increasing very rapidly in NC work.

Three variables which affect the operating conditions of multipoint cutting tools are:

1. *Feed rate*
   The feed should be set on the basis that each cutter tooth remove a specific amount of metal. This is commonly referred to as feed per

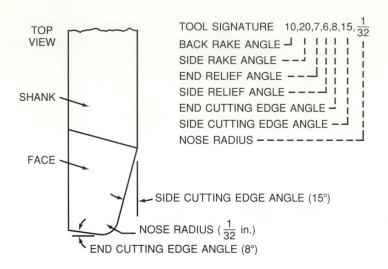

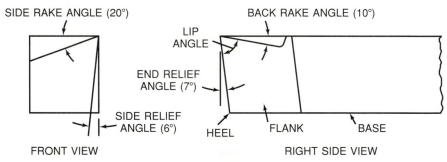

Fig. 8-10 The tool signature code is used to describe single-point tool geometry. (*Allen-Bradley*)

tooth, recommended in feed tables, and should be as close to maximum as possible for each type of cutter. It is a well-proven fact that it is more efficient to remove metal in the form of thick chips instead of thin ones. The maximum feed per tooth is limited by the following factors and must be considered when setting feed rates:

a. Cutting edge strength of the cutter

b. Rigidity of the cutter and the workpiece

c. Surface finish required on the workpiece

d. Tool chip space on a cutter—a two-fluted end mill can take a bigger cut than a four-flute end mill

2. *Cutting speed*

Cutting speed is the rate that the cutting tool revolves and is measured in feet or meters per minute. Recommended cutting speeds for

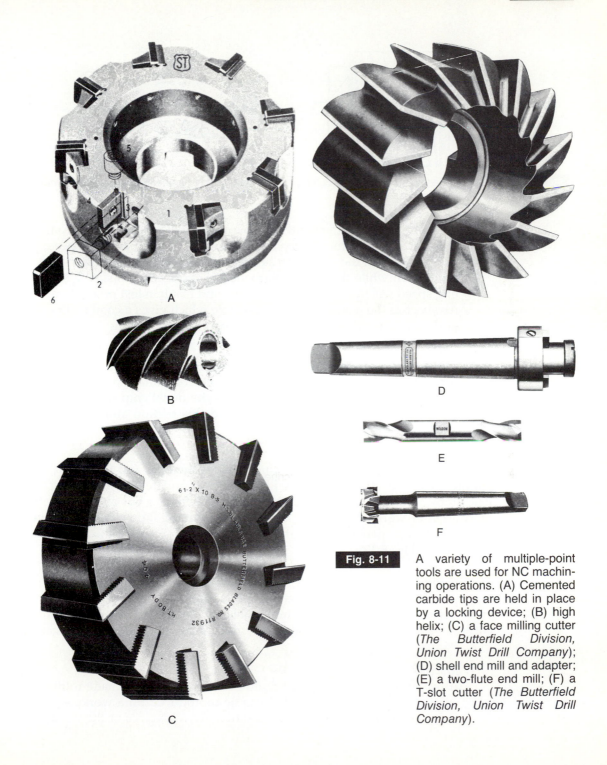

Fig. 8-11 A variety of multiple-point tools are used for NC machining operations. (A) Cemented carbide tips are held in place by a locking device; (B) high helix; (C) a face milling cutter (*The Butterfield Division, Union Twist Drill Company*); (D) shell end mill and adapter; (E) a two-flute end mill; (F) a T-slot cutter (*The Butterfield Division, Union Twist Drill Company*).

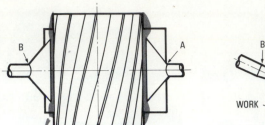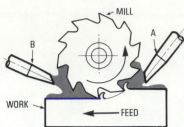

Fig. 8-12 Two nozzles are used to apply cutting fluid for heavy-duty machining operations. (*Cincinnati Milacron, Inc.*)

various types of cutters and workpiece materials are available from manufacturers' or machinery handbooks. Too high a cutting speed will quickly dull the edges of a cutting tool and cause interruptions to production while the tool is sharpened or replaced. Too slow a cutting speed will increase machining time and result in lost production. Always use the maximum speed possible which gives a combination of the best cutting tool life and the best production rates.

3. *Cutting fluid*
   Cutting fluid can play an important role in how efficiently metal is removed in a machining operation. When metal is machined, considerable heat and friction are produced as the metal chip passes over the cutting edge or edges of a tool, and this friction and heat will quickly dull the cutting edges.

Correct application of a cutting fluid reduces the heat and friction at the cutting edge and therefore allows higher speeds and feeds to be used for machining. This not only increases the production rate but also results in more accurate work with a better surface finish. Cutting fluid is most effective when applied in generous amounts at the *exact point* where the cutting is being done (Fig. 8-12).

Special-Purpose Tools   Since NC can be applied economically to almost any type of machine tool, a wide variety of special-purpose cutting tools are generally used. It would be almost impossible to cover every special-purpose cutting tool which is available. However, some of the more common ones are:

1. *Broaches*
   A broach is a special tapered multitoothed cutting tool made to the exact shape and size of the form to be reproduced on a workpiece. Either the broach is moved past a stationary workpiece or the work-

piece is moved past the broach, duplicating the form of the broach into the metal in one pass. Broaching is used extensively for producing intricate internal shapes, cutting slots and keyways, machining flat surfaces on engine blocks or cylinder heads for automobiles, etc.

2. *Superabrasives*

Borazon, cubic boron nitride (CBN), an exciting superabrasive developed by the General Electric Company, is used extensively on NC grinding machines for the grinding of ferrous alloys and superalloys. Borazon crystals, which, after diamond, are the hardest material known, have sharp, long-lasting cutting edges which retain their exceptional strength at very high temperatures.

Borazon CBN grinding wheels grind efficiently at metal removal rates which would cause conventional grinding wheels to dull rapidly. These grinding wheels are ideal for NC grinders because they increase productivity, improve the quality of the finished part, and reduce the amount of downtime required to recondition the grinding wheel.

3. *Taps*

Tapping is the process of producing an internal thread in a workpiece with a cutting tool called a tap. The operation of tapping can be programmed on NC machines. However, the selection of the proper tap is very important.

Taps fall into two categories: hand taps and machine taps. Although hand taps can be used for NC work, machine taps (Fig. 8-13)

Fig. 8-13 Types of machine taps used for NC work. (A) Gun; (B) spiral-fluted; (C) fluteless.

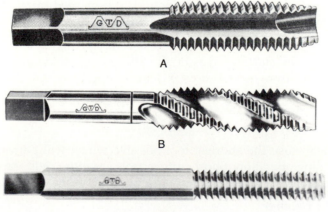

A

B

C

are generally used because of their ability to clear chips quickly. The most common machine taps are:

a. *Spiral point*, sometimes called *gun* or *chip driver taps*. These are recommended for through holes or holes where there is enough clearance in the bottom for the chips.

b. *Spiral-fluted*, generally used when tapping blind holes (those not going through the workpiece). Spiral-fluted taps cut freely, while at the same time clearing chips from the hole by the backward thrust action of the spiral flutes.

c. *Fluteless*. These taps eliminate the problem of clearing chips from a hole because they form or roll threads in a hole. The fluteless tap is actually a forming tool with lobes which is used to produce internal threads in ductile materials such as copper, brass, aluminum, leaded steels, etc.

Cutting Tool Care and Use

Proper selection, care, and use of cutting tools are essential to any machining operation, especially on NC machine tools, because of the accuracy built into these machines. A poor cutting tool or one not suited for a machining operation cannot produce accurate work, and certainly not to the tolerances expected of an NC machine tool. Since there are so many factors which affect machining operations, it would be difficult to give definite solutions to every problem which might arise. However, the points shown in Fig. 8-14 should be considered.

1. *Proper tool selection*
   Choose the cutting tool which will most effectively machine the part to the size and accuracy required. Use cemented carbide and superabrasive cutting tools wherever possible.

2. *Sharp cutting tools*
   Check the cutting edges of all cutting tools; they should be sharp and free from nicks and wear. Dull cutting tools will only get duller, create more heat due to friction, produce poor surface finishes, and result in inaccurate work.

3. *Short rigid tools*
   Always use the shortest tool possible which will perform the machining operation. Short cutting tools are more rigid, have less of a tendency to cause chattering, and produce more accurate work.

4. *Chip clearance*
   Select a cutting tool that provides good chip clearance for the mate-

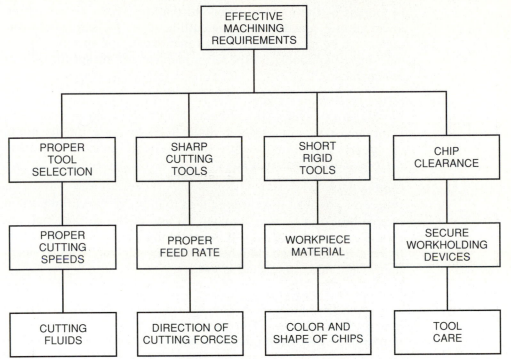

Fig. 8-14 The key factors necessary to machine a part accurately and in the least possible time. (*Modern Machine Shop*)

rial being removed during a cutting operation. Two-fluted end mills are excellent for removing a lot of material quickly, while four- or six-fluted end mills are generally used for finishing purposes.

5. *Proper cutting speeds*
Select the cutting speed for the material being cut that provides the best balance between productivity and cutting tool life. Too high a cutting speed will quickly dull a cutting tool, while too slow a cutting speed will reduce productivity and increase the cost of producing a part.

6. *Proper feed rate*
Use the maximum feed rate possible for the workpiece material, the rigidity of the setup, and the cutting operation being performed. Higher feed rates increase cutter life, while slow feed rates cause a cutter to dull quickly.

7. *Workpiece material*
The cutting speed should be selected in relation to the physical

properties of the material to be cut. Hard, ferrous metals require lower cutting speeds, while the cutting speed for soft, nonferrous metals can be much higher. Check manufacturers' specifications or machinery handbooks for the recommended cutting speeds for various metals.

8. *Secure workholding devices*
Workholding devices should be as simple and rigid as possible to accurately locate and securely hold the workpiece for machining operations. Avoid high fixtures or workholding devices to ensure rigidity and prevent chatter.

9. *Cutting fluids*
Wherever possible, use cutting fluids to improve the cutting action, reduce heat and friction, prolong the life of the cutting tool, and produce a good surface finish. A machining operation using cutting fluids can be operated at higher cutting speeds than a similar operation where no cutting fluid is used.

10. *Direction of cutting forces*
The possibility of vibration or chatter can be greatly reduced or eliminated when the cutting forces are directed towards the solid portion of the workholding device and the machine tool. Always take heavy cuts so that the cutting forces are directed toward the column of the machine or the solid jaw of a vise.

11. *Color and shape of chips*
A change in the shape or color of the chip produced during a machining operation will indicate that the cutting tool is becoming dull. If the chip turns blue when using a high-speed steel cutter, slow the cutting speed to reduce the friction at the cutting edge of the cutter. It is wise to replace or recondition such a cutter as quickly as possible before breakage occurs.

12. *Tool care*
A reasonable amount of care for cutting tools will produce accurate and predictable results for a long time. Never drop or lay any other tools or equipment on a cutting tool, because the cutting edges can become chipped or damaged. Store cutting tools properly (in separate containers, if possible) to protect the cutting edges.

REVIEW QUESTIONS

1. Name three factors that are having a major impact on NC part programming.

NC Programming Languages

2. List five reasons why it is advisable to learn a programming language which is popular.

3. What is APT?

4. Name three characteristics of the APT programming language.

5. Why does the APT processor language give specific and highly predictable results?

6. Briefly describe the NUFORM language.

7. Define SPLIT.

8. Name five major points of the COMPACT II language.

APT General Processor

9. Describe a processor and state what function it serves.

10. Briefly describe the four phases or sections of the APT system.

11. List six main functions of a postprocessor.

APT Programming

12. Name four purposes of APT statements.

13. Define and give an example of explicit and implicit instructions for continuous path motion.

Additional Programming Techniques

14. What is digitizing and how is it used in NC programming?

15. For what purpose was scanning designed?

16. Why is computer graphics valuable to the NC programmer?

NC Tooling Systems

17. What two factors are necessary to obtain the best operating results from an NC machine?

18. Name and state the purpose of the three major categories of cutting tools used on NC machine tools.

19. What type of single-point cutting tools are generally used on turning centers?

20. Name four factors which will limit the maximum feed per tooth or multipoint tools.

21. Why is cutting speed important?

22. How does cutting fluid affect a machining operation?

23. Why are superabrasives such as Borazon CBN finding extensive use on NC grinding machines?

24. Name and state the purpose of three types of taps.

Cutting Tool Care and Use

25. What may result when dull cutting tools are used?

26. Why is chip clearance important?

27. Discuss the effect of cutting speed and feed rate on productivity.

28. Why should cutting forces be directed toward the solid portion of a workholding device or machine tool?

29. What should be done when blue chips are produced by a high-speed steel cutter?

CHAPTER

NINE

Machining Centers

Machining centers evolved from the need to be able to perform a variety of operations and machining sequences on a workpiece on a single machine in one setup. Many parts require machining on several machines and may spend weeks on the shop floor waiting and moving from machine to machine. Studies have shown that a workpiece may spend only 5 percent of its time in the shop on a machine, and only about 30 percent of that 5 percent, or 1.5 percent, in actual machining time.

Operations such as milling, contouring, drilling, counterboring, boring, spotfacing, and tapping can now be performed on machining centers in any sequence and require only one setup. Machining centers equipped with automatic tool changers, rotary tables, and rotary work heads make this a very versatile machine while reducing the operator intervention during the cutting cycle.

After completing
this chapter,
you should be
able to:

1. Understand the purpose and function of machining centers

2. Explain the application of computerized numerical control (CNC) for machining centers

3. Identify the types of machining operations for which the machine was designed

TYPES OF MACHINING CENTERS

There are two main types of machining centers: the horizontal spindle and the vertical spindle machine.

Horizontal Spindle Type

1. The traveling-column type (Fig. 9-1) is equipped with one or usually two tables on which the work can be mounted. With this type of machining center, the workpiece can be machined while the operator is loading a new workpiece on the other table.

2. The fixed-column type (Fig. 9-2) is equipped with a pallet shuttle. The pallet is a removable table on which the workpiece is mounted. After the workpiece has been machined, the workpiece and pallet are moved to a shuttle which then rotates, bringing a new pallet and workpiece into position for machining.

Vertical Spindle Type

The vertical spindle machining center (Fig. 9-3) is a saddle-type construction with sliding bedways which utilizes a sliding vertical head instead of a quill movement.

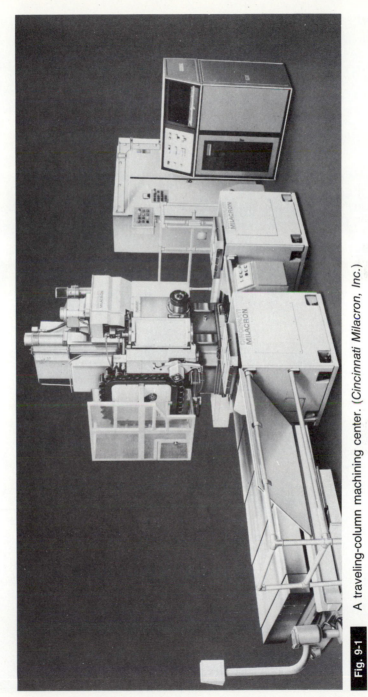

Fig. 9-1 A traveling-column machining center. *(Cincinnati Milacron, Inc.)*

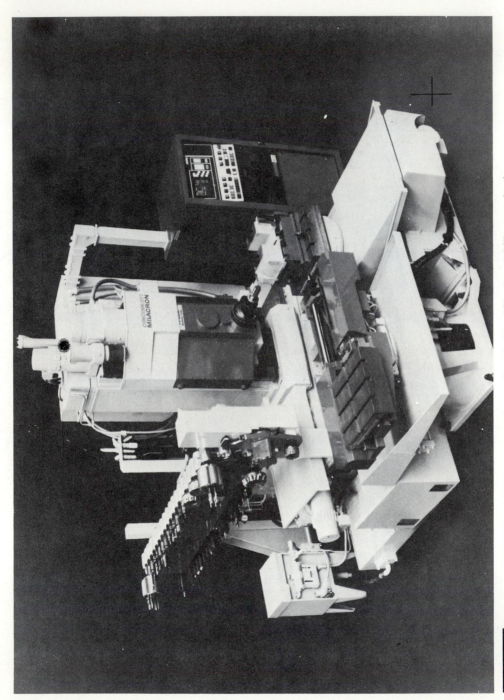

Fig. 9-2 A fixed-column machining center with a pallet shuttle. (*Cincinnati Milacron, Inc.*)

Fig. 9-3 A vertical CNC machining center with a tool change storage drum. (*Bridgeport Machines, Inc.*)

PARTS OF THE CNC MACHINING CENTERS

The main parts of CNC machining centers are the bed, saddle, column, table, servo motors, ball screws, spindle, tool changer, and the machine control unit (MCU) (Fig. 9-4).

> *Bed*—The bed is usually made of high-quality cast iron which provides for a rigid machine capable of performing heavy-duty machining and maintaining high precision. Hardened and ground ways are mounted to the bed to provide rigid support for all linear axes.
>
> *Saddle*—The saddle, which is mounted on the hardened and ground bedways, provides the machining center with the X-axis linear movement.
>
> *Column*—The column, which is mounted to the saddle, is designed

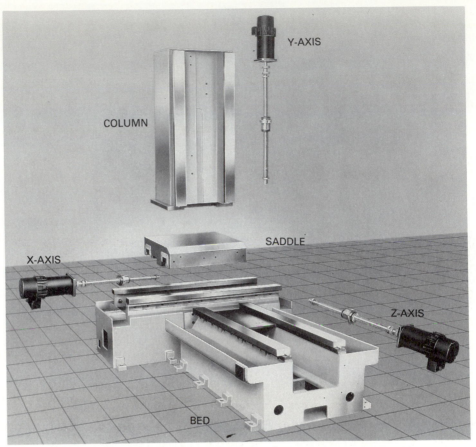

The main parts of a CNC machining center. (*Cincinnati Milacron, Inc.*)

with high torsional strength to prevent distortion and deflection during machining. The column provides the machining center with the Y-axis linear movement.

Table—The table, which is mounted on the bed (Fig. 9-5), provides the machining center with the Z-axis linear movement.

Servo system—The servo system, which consists of servo drive motors, ball screws, and position feedback encoders, provides fast, accurate movement and positioning of the XYZ axes slides. The feedback encoders mounted on the ends of the ball screws form a closed-loop system which maintains consistent high-positioning unidirection repeatability of ±0.0001 in. (0.003 mm).

Spindle—The spindle, which is programmable in 1-r/min increments, has a speed range of from 20 to 6000 r/min. The spindle can be of a fixed position (horizontal) type, or can be a tilting/contouring spindle which provides for an additional (A) axis (Fig. 9-6).

Tool changers—There are basically two types of tool changers, the vertical tool changer (Fig. 9-7) and the horizontal tool changer (Fig. 9-8). The tool changer is capable of storing a number of preset tools which can be automatically called for use by the part program. Tool changers are usually bidirectional, which allows for the shortest travel distance to randomly access a tool. The actual tool change time is usually only 3 to 5 s.

MCU—The MCU allows the operator to perform a variety of operations such as programming, machining, diagnostics, tool and machine monitoring, etc. MCUs vary according to manufacturers' specifications; new MCUs are becoming more sophisticated, making machine tools more reliable and the entire machining operations less dependent on human skills.

Fig. 9-5 The table provides the machining center with the Z axis linear movement. (*Cincinnati Milacron, Inc.*)

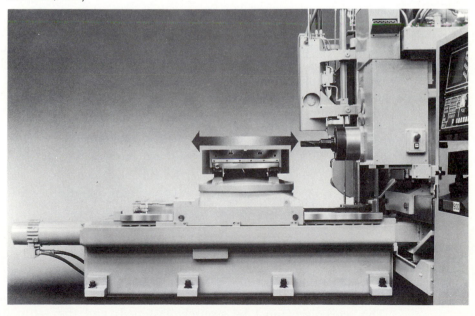

Fig. 9-6   The tilting contouring spindle provides an additional A axis. (*Cincinnati Milacron, Inc.*)

MACHINE AXES

Machining centers have probably made the greatest impact in NC machining because of their ability to perform such a variety of machining operations on all sides of a workpiece with only one setup. The five-axis machining center (Fig. 9-9) indicates the axes that can be used when performing these machining operations and sequences.

X axis	Linear movement
Y axis	Linear movement
Z axis	Linear movement
A axis	Tilt/contour spindle
B axis	Rotary table

WORKHOLDING DEVICES

When a workpiece is set up, it is important to ensure that the setup be safe. The workpiece must be securely fastened, and the setup must be rigid

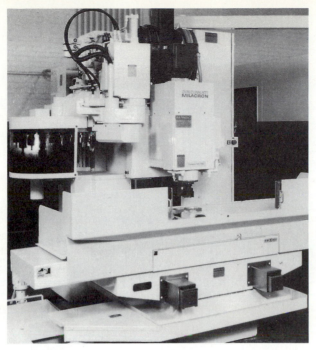

Fig. 9-7 The vertical tool changer is capable of automatically changing tools by tape or computer command. (*Cincinnati Milacron, Inc.*)

Fig. 9-8 The horizontal tool changer is capable of automatically changing tools by tape or computer command. (*Cincinnati Milacron, Inc.*)

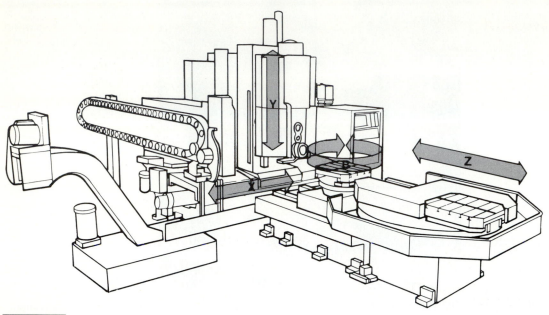

Fig. 9-9   A five-axis machining center can perform a wide variety of machining operations. (*Cincinnati Milacron, Inc.*)

enough to withstand the forces which will be present during the machining operation. If the workpiece or the holding device becomes loose during machining, damage can result to the tooling and/or the machine.

The machine operator should be sure that all workholding devices are free from chips and burrs before using. The workholding devices, generally specified by the programmer, should be located in the proper position on the machine table. Failure to follow these instructions may result in operator injury, damage to the machine, or scrapped workpieces.

Types of Workholding Devices

Swivel-base vise (Fig. 9-10A)—May be bolted to the table or subplate. The swivel base enables the vise to be swivelled 360° in a horizontal plane.

Angle plates (Fig. 9-10B)—Are L-shaped pieces of cast iron or steel accurately machined to a 90° angle. They are made in a variety of sizes and have holes or slots which provide a means for fastening the workpiece.

V blocks (Fig. 9-10C)—Are generally used in pairs to support round

Fig. 9-10   (A) The swivel-base vise can be swiveled through 360° in a horizontal plane. (B) Angle plates are machined to an accurate 90° angle (*Kostel Enterprises Ltd.*); (C) V blocks are used to hold round work for machining (*L. S. Starrett Co.*).

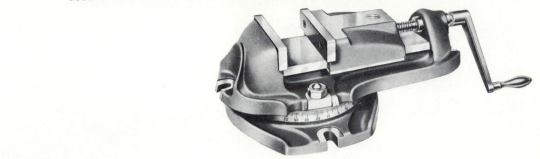

A

B

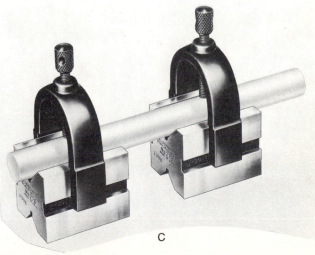

C

Fig. 9-11 (A) Step blocks support the end of the clamp (*Northwestern Tools, Inc.*). (B) Clamps and straps are used to fasten work to the machine table (*J. W. Williams & Co.*). (C) Support jacks are used to support and prevent the work from distorting when being clamped (*Kosel Enterprises Ltd.*). (D) Parallels are used to support the workpiece.

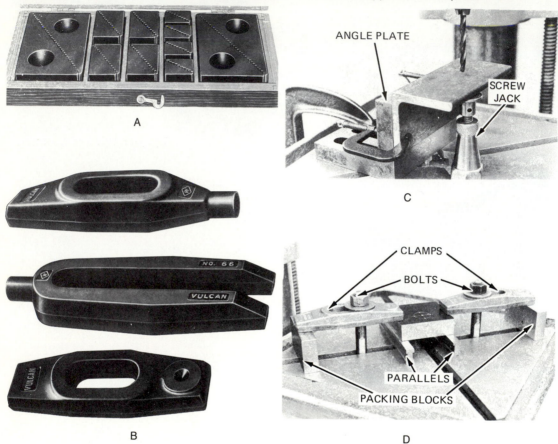

work. A U-shaped clamp may be used to fasten the work in a V block.

Step blocks (Fig. 9-11A)—Are used to provide support for strap clamps when work is being fastened to the table or workholding device.

Clamps or *Straps* (Fig. 9-11B)—Are used to fasten work to the table, angle plate, or fixture. They are made in a variety of sizes and are usually supported at the end by a step block and bolted to the table by a T bolt. It is good practice to place the T bolt in the clamp or strap as close to the work as possible.

Support jacks (Fig. 9-11C)—Are used to support the workpiece to prevent distortion of the workpiece during clamping.

Parallels (Fig. 9-11D)—Are flat, square, or rectangular pieces of metal used to support the workpiece for setup.

Subplates—Are generally flat plates that may be fitted to the machine table to provide quick and accurate location of workpieces, workholding devices, or fixtures. The fixturing holes in these subplates are accurately located and, when set up on the machine table in relation to the machine datum, provide the programmer with known locating positions.

FIXTURES

NC eliminates many of the expensive jigs and fixtures which were previously necessary to hold and locate a workpiece on conventional machine tools. The repetitive position accuracy of an NC machine tool eliminates the need for guide bushings which were previously required to locate the cutting tool.

NC fixtures (Fig. 9-12) are used to accurately locate a part and hold it securely for any machining operations required. Fixture design should be kept as simple as possible so that the time required to load and unload a part is kept as short as possible. Since this time is nonproductive time, the savings here will result in corresponding savings in the cost of producing a part. When designing a fixture to hold a part, you must consider the following points:

1. *Positive location:* The fixture must hold a workpiece securely enough to prevent the workpiece from linear movement in the X, Y, and Z axes, and rotational movement in either direction about each axis.

2. *Repeatability:* Identical parts should always be located in exactly the same location for every part change.

3. *Ruggedness:* Fixtures must be designed to withstand the shock occurring during the machining and loading and unloading cycles.

4. *Rigidity:* The workpiece must be held securely enough to prevent any movement due to the forces created by the machining operation.

5. *Design:* Modular fixtures using standard components are quicker to produce and less costly than custom fixtures. They can also be quickly modified to accommodate differently shaped parts.

6. *Low profile:* Parts of the fixture or the necessary clamping devices to

Fig. 9-12   NC fixtures are used to accurately locate a part and hold it securely for machining operations. (*Cincinnati Milacron, Inc.*)

hold the part should be designed so that there is free movement for the cutting tool at any point in the machining cycle.

7. *Part loading/unloading:* The fixture and its clamping devices should be designed so that they do not interfere with the rapid loading or unloading of a part.

8. *Part distortion:* The fixture should be designed so that the part being machined is not distorted by gravity, machining forces, or clamping forces. Stress should never be put on a part by the clamping forces; otherwise the machined part will distort when the clamping forces are removed.

Clamping Hints

1. Always place the bolt as close to the work as possible.

2. Place a piece of soft metal ("packing") between the clamp and the

workpiece to prevent damage to the workpiece and to spread the clamping force over a wider area.

3. Make sure the packing is not extending into the machining path of the cutting tool.

4. Use the table slots to prevent round work from moving.

5. Use two clamps whenever possible.

6. Parts that do not lie flat should be shimmed to prevent the work from rocking. This will also prevent distortion when the work is clamped.

7. Tighten clamping bolts evenly to prevent workpiece distortion.

SPEEDS AND FEEDS

The most important factors affecting the efficiency of machining on a machining center are cutter speed, feed, and depth of cut. If a cutting tool is run too slowly, valuable time will be wasted, resulting in lost production. Too high a speed will create too much heat and friction at the cutting edge of the tool, which quickly dulls the cutter. This results in having to stop the machine to either recondition or replace the cutter. Somewhere between these two extremes is the efficient cutting speed for each material being cut.

Feed is the rate at which the work is fed into a revolving cutter. If the work is fed too slowly, time will be wasted, resulting in lost production, and cutter chatter, which shortens the life of the cutter, may occur. If work is fed too fast, the cutter teeth can be broken. Much time will be wasted if several shallow cuts are taken instead of one deep cut or roughing cut. Therefore, speed, feed, and depth of cut are three important factors which affect the life of the cutting tool and the productivity of the machine tool. Charts and tables containing speeds, feeds, and depth of cut are available from cutting tool manufacturers, machinery handbooks, etc.

CUTTING TOOLS

The selection of the proper cutting tools for each operation on a machining center is essential to producing an accurate part. Generally there is not enough thought and planning going into the selection of cutting tools for each particular job. The NC programmer must have a thorough knowledge of cutting tools and their applications in order to properly program any part.

Machining centers use a variety of cutting tools to perform various machining operations. These tools may be conventional high-speed steel, cemented carbide inserts, CBN (cubic boron nitride) inserts, or polycrystalline diamond insert tools. Some of the tools used are end mills, drills, taps, reamers, boring tools, etc.

Studies show that machining center time consists of 20 percent milling, 10 percent boring, and 70 percent hole-making in an average machine cycle. On conventional milling machines, the cutting tool cuts approximately 20 percent of the time, while on machining centers the cutting time can be as high as 75 percent. The end result is that there is a larger consumption of disposable tools due to decreased tool life through increased tool usage.

End mills
End mills and shell end mills (Fig. 9-13) are widely used in machining centers. They are capable of performing a variety of machining operations such as face, pocket, and contour milling; spotfacing; counterboring; and roughing and finishing of holes using circular interpolation.

Drills
Conventional as well as special drills are used to produce holes (Fig. 9-14). Always choose the shortest drill that will produce a hole of the required depth. As drill diameter and length increase, so does the error in hole size and location. Stub drills are recommended for drilling on machining centers.

Center drills
Center drills (Fig. 9-15) are used to provide an accurate hole location for the drill which is to follow. The disadvantage of using center drills is that the small pilot drill can break easily unless care is used. An alternative to the center drill is the spotting tool, which has a 90° included angle and is widely used for spotting hole locations.

Taps
Machine taps (Fig. 9-16) are designed to withstand the torque required to thread a hole and clear the chips out of the hole. Tapping is one of the most difficult machining operations to perform because of the following factors:

- Inadequate chip clearance
- Inadequate supply of cutting fluid
- Coarse and fine threads in various materials
- Speed and feed of threading operations being governed by the lead of the thread
- Depth of thread required

Fig. 9-13 End mills are used for a wide variety of machining operations. (A) Shell end mill and adapter (*Cleveland Twist Drill*); (B) two-flute end mill (*Weldon*); (C) four-flute end mill (*Butterfield*).

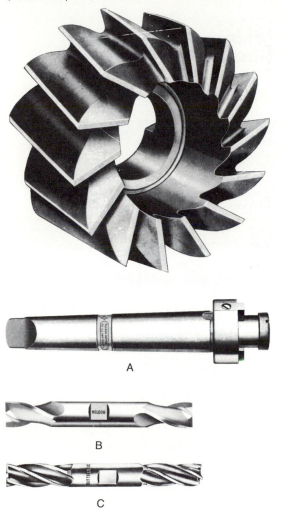

A

B

C

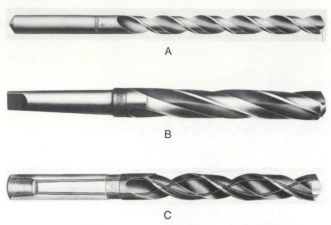

Fig. 9-14 Stub drills are widely used on machining centers. (A) A high helix drill; (B) a core drill; (C) an oil hole drill. (*Cleveland Twist Drill Company*)

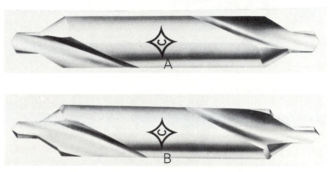

Fig. 9-15 Center drills are generally used to accurately spot the location of holes. (A) Regular type; (B) bell type. (*Cleveland Twist Drill Company*)

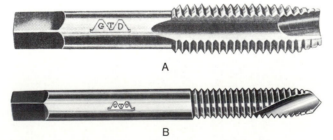

Fig. 9-16 Taps are used to produce a variety of internal threads. (A) Gun; (B) stub flute. (*Continued on page 233.*)

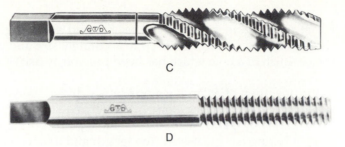

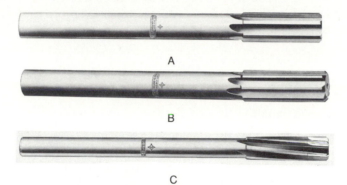

Fig. 9-16   (*Continued*) (C) spiral flute; (D) fluteless. (*Greenfield Tap & Die*)

Fig. 9-17   Reamers are used to accurately size a hole and produce a good surface finish. (A) Rose reamer; (B) fluted reamer; (C) carbide-tipped reamer. (*Cleveland Twist Drill Company*)

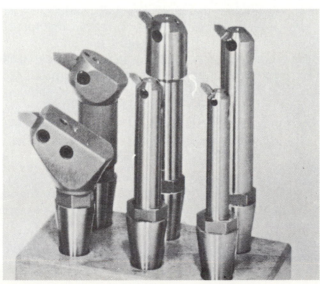

Fig. 9-18   Single-point boring tools are used to enlarge a hole and bring it to location. (*Moore Special Tool Co.*)

Reamers

Reamers are available in a variety of designs and sizes (Fig. 9-17). A reamer is a rotary end cutting tool used to accurately size and produce a good surface finish in a hole which has been previously drilled or bored.

Boring tools

Boring is the operation of enlarging a previously drilled, bored, or cored hole to an accurate size and location with a desired surface finish. This operation is generally performed with a single-point boring tool (Fig. 9-18). When a boring bar is selected, the length and diameter should be carefully considered: as the ratio between length and diameter increases, the rigidity of the boring bar decreases. For example, a boring bar with a 1:1 length-to-diameter ratio is 64 times more rigid than one with 4:1 ratio.

TOOLING SYSTEMS

Toolholders

The machining center, a multifunction machine tool, uses a wide variety of cutting tools such as drills, taps, reamers, end mills, face mills, boring tools, etc., to perform various machining operations on a workpiece. For these cutting tools to be inserted into the machine spindle quickly and accurately, all these tools must have the same taper shank toolholders to suit the machine spindle. The most common taper used in NC machining center spindles is the No. 50 taper, which is a self-releasing taper. The toolholder must also have a flange or collar, for the tool-change arm to grab, and a stud, tapped hole, or some other device for holding the tool securely in the spindle by a power drawbar or other holding mechanism (Fig. 9-19A and B).

When one is preparing for a machining sequence, the tool assembly drawing is used to select all the cutting tools required to machine the part. Each cutting tool is then assembled off-line in a suitable toolholder and preset to the correct length. Once all the cutting tools are assembled and preset, they are loaded into specific pocket locations in the machine's tool-storage magazine where they are automatically selected as required by the part program.

Tool Identification

NC machine tools use a variety of methods to identify the various cutting tools which are used for machining operations. The most common methods of identifying tools are:

1. *Tool pocket locations*
   Tools for early machining centers were assigned a specific pocket

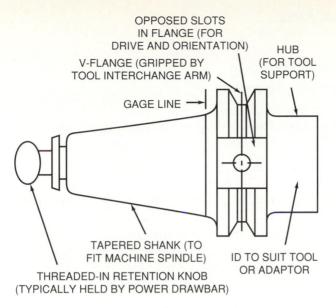

OPPOSED SLOTS
IN FLANGE (FOR
DRIVE AND ORIENTATION)

V-FLANGE (GRIPPED BY
TOOL INTERCHANGE ARM)

HUB
(FOR TOOL
SUPPORT)

GAGE LINE →

TAPERED SHANK (TO
FIT MACHINE SPINDLE)

THREADED-IN RETENTION KNOB
(TYPICALLY HELD BY POWER DRAWBAR)

ID TO SUIT TOOL
OR ADAPTOR

A

B

Fig. 9-19 NC toolholders are rugged precision tools which are designed to be located accurately in a machine spindle by an automatic tool-changing system. (*Hertel Carbide Ltd.*)

location in the tool-storage magazine, and each tool was called up for use by the part program.

2. *Coded rings on toolholders*
   A special interchange device reader was used to identify some tools by special coded rings on the toolholder.

3. *Tool assembly number*
   Most modern MCUs have a tool identification feature which allows the part program to recall a tool from the tool-storage magazine pocket by using a five- to eight-digit tool assembly number.
   - Each tool assembly number may be assigned a specific pocket in the tool-storage magazine by the tool data tape, by the operator using the MCU, or by a remote tool management console.

Tool Management Program

It is important, to achieve the best productivity from any machine tool, to have a sound program which covers all aspects of cutting tools. The best NC machine tool cannot come anywhere close to its productivity potential unless the best cutting tools for each operation are available for use when they are required. The tool-management program must include such things as tool design, standard coding system, purchasing, good tooling practices, part programming which is cost effective, and the best use of cutting tools on the machine.

A good cutting tool policy must include the following:

1. *Standard policy*
   - A standard policy regarding cutting tools must be established.
   - Everyone should clearly understand the policy: the tool engineer, the part programmer, the supervisory staff, the setup person, and the machine tool operator.
   - The role that each person has in selecting the proper cutting tools must be clearly defined.

2. *Cutting tool dimensional standards*
   - All cutting tools purchased or specially made must conform to established cutting tool dimensional standards.
   - When it is necessary to recondition cutting tools, they should be ground to the next NC standard.
   - The part programmer must use the cutting tool standards for programming purposes.

3. *Rigid cutting tools*
   - Always select the shortest cutting tool possible for each job to ensure locational accuracy and rigidity.
   - Cutting tool holders should be of one-piece construction to provide rigidity.

4. *Tool preparation*
   - There must be a rigid policy on tool setting, compensation, and regrinding which is understood by everyone concerned.
   - Clearly define who has the responsibility for each so that there is no conflict or misunderstanding.

5. *Indexable insert tools*
   - Use cemented carbide insert-type tooling wherever possible because of their wear resistance, higher productivity, and dimensional accuracy.
   - Borazon (CBN) inserts should be used on hard ferrous metals where cemented carbides are not satisfactory.
   - Synthetic (polycrystalline) diamond inserts should be used for machining nonferrous materials.

The success or failure of a tool-management program depends in large part on the part programmer. To be most effective, the programmer must have a thorough knowledge of machining practices and procedures and the type of cutting tool required for each operation. Most modern CNC units have standard or optional features or programs available to make any tool-management policy more effective.

RANDOM AND SEQUENTIAL TOOLING SYSTEMS

Most modern NC machine tools are equipped with automatic tool changers to quickly replace cutting tools for the next machining operation. These tool changers may be equipped for either random or sequential tooling selection.

Random tooling selection is a system where there is no specific pattern of tool selection. Random tooling is generally in wider use in industry because of the flexibility it offers over sequential tooling. With random tooling, each tool is given a specific tool identification number and is loaded into a specific pocket in the tool magazine. As each tool is required for use in the NC program, the previous tool is removed from the machine tool spindle by the tool changer arm and replaced in the correct tool-magazine pocket. The new tool, as selected by the NC program, is taken from the correct tool-magazine pocket and inserted into the machine spindle. Whenever a certain cutting tool or one that has been used before is needed for machining, the MCU knows where to find it.

Sequential tooling selection is a system where all tools must be loaded in the exact order in which they will be used to machine a part. Therefore it is very important that the correct order of tooling be programmed and loaded in the tool magazine in the order they are required to complete the machining operations on a part. If the cutting tools are not in the correct order, the next tool is automatically selected and the machine may try to tap a hole with an end mill. Therefore, if it is necessary to use a tool more than once, it is necessary to load similar tools in the tool magazine in the order that they are used.

ADAPTIVE CONTROL

The programmer must select the proper speeds and feeds for each part to be machined, so that the part is produced in the shortest period of time while considering cutting tool life. Generally, cutting tool breakage occurs because the cutter is dull, or the depth of cut is changed because of variances in workpiece thickness. Whenever a cutting tool becomes dull or is broken, the NC machine must be stopped to recondition or replace the cutting tool. To get the best productivity from an NC machine tool, optimum speeds and feeds should be used for machining operations.

A feature that is fast becoming popular is that of *torque control machining*, the torque being calculated from measurements at the spindle drive motor (Fig. 9-20). This device will increase productivity by preventing or sensing damage to the cutting tool. The torque is measured when the machine is turning but not cutting, and this value is stored in the computer memory.

As the machining operation begins, the stored value is subtracted from the torque reading at the motor. This will give the net cutting torque, which is compared to the programmed torque or limits stored in the computer (or on NC tape). If the net cutting torque exceeds the programmed torque limits, the computer will act by reducing the feed rate, turning on the coolant, or even stopping the cycle. The feed rate will be lowered whenever the horsepower requirements exceed the rated motor capacity or the programmed code value.

The system display of three yellow lights advises the operator of the operational conditions in the machine at the time. A left-hand yellow light indicates that the torque control unit is in operation. The middle yellow light indicates that the horsepower limits are being exceeded. The right-hand light comes on

Fig. 9-20 Torque control increases or decreases the feed rate depending upon the depth of cut or the dullness of the cutting tool at any time during a machining cycle. (*Cincinnati Milacron, Inc.*)

when the feed rate drops below 60 percent of the programmed rate. The meter (Fig. 9-20) indicates the cutting torque (or operational feed rate) as a percent of the programmed feed rate.

As the tool gets dull, the torque will increase and the machine will back off on the feed and ascertain the problem. Excessive material could be on the workpiece, or a tool might be very dull or broken. If the tool is dull, the machine will finish the operation and a new backup tool of the same size will be selected from the storage chain when that operation is performed again. If the torque is too great, the machine will stop the operation on the workpiece and program the next piece into position for machining.

PREPARATORY FUNCTIONS FOR MACHINING CENTERS—G CODES

Preparatory function codes, or G codes, are the heart of an NC program. As the name implies, it prepares the machine so it knows what to do with the rest of the information.

G codes are divided into different groups from 00 to 10. If G codes of the same group are used in the same block, only the last G code in the block becomes effective. G codes in groups 1 to 10 are *modal*, which means they remain in memory. Once used, they do not have to be repeated in every consecutive block. The G codes in group 00 are *nonmodal* and have to be repeated in every block where they are used.

Group	*G code*	*Function*
01	G00	Rapid positioning
01	G01	Linear interpolation

Group	G code	Function
01	G02	Circular interpolation clockwise (CW)
01	G03	Circular interpolation counterclockwise (CCW)
00	G04	Dwell
00	G10	Offset value setting
02	G17	XY plane selection
02	G18	ZX plane selection
02	G19	YZ plane selection
06	G20	Inch input (in.)
06	G21	Metric input (mm)
04	G22	Stored stroke limit on
04	G23	Stored stroke limit off
00	G27	Reference point return check
00	G28	Return to reference point
00	G29	Return from reference point
00	G30	Return to second reference point
07	G40	Cutter compensation cancel
07	G41	Cutter compensation left
07	G42	Cutter compensation right
08	G43	Tool length compensation in positive (+) direction
08	G44	Tool length compensation in minus (−) direction
08	G49	Tool length compensation cancel
00	G45	Tool offset expansion
00	G46	Tool offset reduction
00	G47	Tool offset double expansion
00	G48	Tool offset double reduction
09	G74	Left-hand tapping cycle
09	G76	Fine boring
09	G80	Canned cycle cancel
09	G81	Drill cycle, spot boring
09	G82	Drilling cycle, counterboring
09	G83	Peck drilling cycle
09	G84	Tapping cycle
09	G85	Boring cycle #1
09	G86	Boring cycle #2

Group	G code	Function
09	G87	Boring cycle #3
09	G88	Boring cycle #4
09	G89	Boring cycle #5
03	G90	Absolute programming
03	G91	Incremental programming
00	G92	Setting of program zero point
05	G94	Feed per minute
10	G98	Return to initial point in canned cycle
10	G99	Return to R point in canned cycle

MISCELLANEOUS FUNCTIONS—M CODES

The M code command is used to control miscellaneous functions such as spindle start, stop, tool change, etc. Only one M function can be specified in one block of information. All of the M functions are as follows:

Code	Function
M00	Program stop
M01	Optional stop
M02	End of program
M03	Spindle start (forward CW)
M04	Spindle start (reverse CCW)
M05	Spindle stop
M06	Tool change
M07	Mist coolant on
M08	Flood coolant on
M09	Coolant off
M19	Spindle orientation
M30	End of tape (return to top of memory)
M48	Override cancel release
M49	Override cancel
M98	Transfer to subprogram
M99	Transfer to main program (subprogram end)

REVIEW QUESTIONS

Machining Centers

1. List six operations that could be performed on a machining center.

2. Name three accessories that make machining centers so versatile.

Types of Machining Centers

3. What are two types of machining centers?

4. Briefly describe the difference between a traveling-column and a fixed-column machining center.

5. How does a vertical machining center provide for Z-axis movement?

Parts of the CNC Machining Center

6. Name and state the purpose of three main parts of a machining center.

7. In which direction is linear motion provided by the:
   (a) Column?
   (b) Table?

8. What is the purpose of the feedback encoders?

9. What is the positioning repeatability of the machining center?

10. What is meant by a "closed-loop system"?

11. What is the smallest programmable increment for the spindle?

12. Name and state the difference between two types of spindles used on machining centers.

13. Name and state two types of tool changers used on machining centers.

14. What is meant by the term *bidirectional tool changer?*

15. How long does it take to make a tool change?

16. What is the purpose of an MCU?

Machine Axes

17. Why have machining centers made such an impact on NC machining?

18. List the five axes usually found on a five-axis machining center.

Workholding Devices

19. What are three important factors to consider when setting up a workpiece?

20. Why should workholding devices be placed on the work table as specified by the programmer and not the setup person?

21. Name and state the purpose of seven workholding devices used on machining centers.

22. List eight things that should be taken into consideration when designing a fixture.

Clamping Hints

23. Why should the clamping bolts be placed as close to the work as possible?

24. What is the purpose of using packing when clamping a workpiece?

25. How can the table slots be used to prevent round work from moving?

26. Why is it important to tighten clamping bolts evenly?

Speeds and Feeds

27. What three factors affect the efficiency of a machining center?

28. List two problems that may arise from a cutting tool running too slow or too fast.

29. Briefly define the term *feed*.

30. List two problems that may arise from a feed rate that is too slow or too fast.

31. Name three factors that offset the life of a cutting tool.

Cutting Tools

32. Why must a programmer have a thorough knowledge of cutting tools and their applications?

33. List five tools that are commonly used on machining centers.

34. List seven operations that can be performed on a machining center using end mills.

35. Why are stub drills recommended when drilling on a machining center?

36. What is the purpose of a center drill?

37. Why is tapping considered to be a difficult machining operation to perform?

38. What is the purpose of a reamer?

39. Briefly define the term *boring*.

40. Why is it important to consider the length and diameter of the boring bar?

Tool Identification

41. Name and briefly describe three methods of identifying tools.

Tool-Management Program

42. Name and state the purpose of the five factors of a good cutting tool policy.

43. What knowledge must an effective part programmer have?

Random and Sequential Tooling Systems

44. Define *random tooling*.

45. How does the MCU know where each tool is in the tool-storage magazine?

46. Define *sequential tooling*.

47. Why is it important that all cutting tools be stored in the tool magazine in the proper sequence?

Adaptive Control

48. Name two reasons why cutting tool breakage generally occurs.

49. What is torque control machining and how does it operate?

Preparatory Functions

50. What is the purpose of G codes?

51. Explain the term *modal*.

52. What functions do the following G codes perform?
   (a) G00
   (b) G03
   (c) G28
   (d) G40

(e) G83

(f) G90

Miscellaneous Functions

53. What is the purpose of M codes?

54. What functions do the following M codes perform?

(a) M00

(b) M02

(c) M03

(d) M06

(e) M09

(f) M30

CHAPTER

TEN

Chucking and Turning Centers

Numerical controlled (NC) lathes had a very slow start in manufacturing. Studies during the 1960s indicated that 40 percent of all metal-cutting operations were performed on lathes, yet NC lathes accounted for only 7.4 percent of the total NC sales at that time.

These lathes were standard engine lathes that had been retrofitted for NC. The term *retrofit* means fitting control systems and other required NC gear to a conventional piece of equipment that was not originally designed for NC. These lathes were a significant improvement over the conventional and tracer lathes being used at that time. They were capable of making contour cuts by controlling the coordinated motion of the cross-slide and the carriage. Thread-cutting was made possible by the automatic synchronization of the spindle revolutions per minute (r/min) and the travel of the carriage.

Through continued research and development, today's computerized numerical controlled (CNC) chucking and turning centers are capable of greater precision and higher production rates. These machine tools can be equipped with color graphics display screens for programming, in-process gaging, tool changers, machining monitors, rotary tooling, and automatic loading and unloading devices. This allows the machine to run virtually unattended with a minimum of downtime, making CNC chucking and turning centers exceptionally versatile machine tools.

Objectives

After completing
this chapter,
you should be
able to:

1. Describe the purpose and functions of chucking and turning centers

2. List the applications of CNC for chucking and turning centers

3. Identify the types of machining operations for which each machine is designed

TYPES OF TURNING CENTERS

CNC Chucking Center

The CNC chucking center (Fig. 10-1) is designed to machine most work that is held in a chuck. These machines are manufactured in a wide variety of sizes, with chuck sizes ranging from 8 to 36 in. (200 to 900 mm) in diameter, spindle drive motors from 5 to 50 horsepower, and spindle speeds up to 5000 r/min.

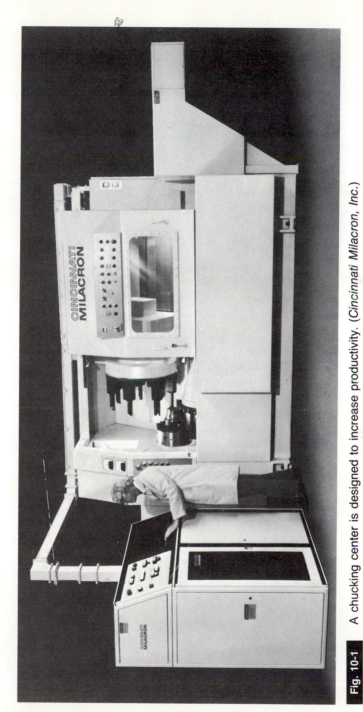

Fig. 10-1 A chucking center is designed to increase productivity. (*Cincinnati Milacron, Inc.*)

Fig. 10-2 Two turrets simultaneously machining the outside of a workpiece. (*Cincinnati Milacron, Inc.*)

The four-axis chucking center incorporates two turrets operating independently on separate slides, machining the workpiece simultaneously. While the upper turret is machining the inside diameter, the lower turret may be machining the outside diameter (Figs. 10-2 and 10-3). However, if the workpiece requires mainly internal operations, both turrets can work on the inside of the workpiece simultaneously (Fig. 10-4). This type of operation is suitable for large-diameter parts which require boring, chamfering, threading, internal radii, or retaining grooves.

For parts with mostly outside-diameter operations, the upper turret can be equipped with turning tools so that both turrets can machine the outside diameter. Figure 10-5 shows a diameter and a chamfer being machined at the same time.

When longer parts must be machined, the right-hand end of the shaft may be supported with a center mounted in the upper turret while the lower turret performs the external machining operations (Fig. 10-5). Other operations which may be performed simultaneously are turning and facing (Fig. 10-6) and internal and external threading (Fig. 10-7).

Other chucking centers are chiefly two-axis models. They may have a single disklike turret on which both inside-diameter and outside-diameter tools are mounted, or they may have two turrets (usually on the same slide). In the latter case, one turret is normally designed for outside-diameter tools and the

Fig. 10-3   Turning and boring operations may be carried on simultaneously. (*Cincinnati Milacron, Inc.*)

Fig. 10-4   Both turrets being used for internal machining operations. (*Cincinnati Milacron, Inc.*)

Fig. 10-5 The upper turret supports a long shaft while the lower turret performs the machining operation. (*Cincinnati Milacron, Inc.*)

Fig. 10-6 Turning and facing operations may be performed by using two turrets. (*Cincinnati Milacron, Inc.*)

Fig. 10-7   Cutting internal and external threads simultaneously. (*Cincinnati Milacron, Inc.*)

other for inside-diameter tools. Whatever the arrangement, the two-axis control will drive only one turret at a time.

CNC Turning Center

CNC turning centers (Fig. 10-8), while not unlike chucking centers, are designed mainly for machining shaft-type workpieces which are supported by a chuck and a heavy-duty tailstock center.

On four-axis machines two opposed turrets, each capable of holding different tools, are mounted on separate cross-slides, one above and one below the centerline of the work. Because the turrets balance the cutting forces applied to the work, extremely heavy cuts can be taken on a workpiece when it is supported by the tailstock. The dual turrets also lend themselves to other operations, such as:

- Roughing and finishing cuts in one pass
- Machining different diameters on a shaft simultaneously (Fig. 10-9)
- Finish turning and threading simultaneously
- Cutting two different sections of a shaft at the same time

When the turning center is equipped with a steadyrest, operations such as facing and threading may be performed on the end of a shaft (Fig. 10-10).

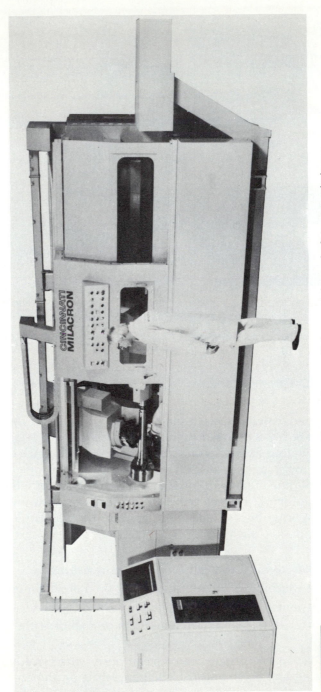

Fig. 10-8 CNC turning center is designed for maximum efficiency on long shaft-type workpieces. (*Cincinnati Milacron, Inc.*)

Fig. 10-9   Both turrets being used to machine a shaft with different diameters. (*Cincinnati Milacron, Inc.*)

Fig. 10-10   A steadyrest supporting a workpiece so that machining may be performed on the end of the shaft. (*Cincinnati Milacron, Inc.*)

CNC CENTER-DRIVE LATHES

The NC center-drive lathe (Fig. 10-11) makes it possible to machine parts from both sides without relocating the workpiece. Machining operations such as turning, facing, boring, etc., can be performed either individually or simultaneously on each side of the workpiece.

The center-drive headstock, in which the workpiece is mounted, is located at the center of the machine and is straddled by two opposing multitool turrets which can perform a wide variety of machining operations. The tool turrets are hydraulically indexed in either direction, and the turret carriages are individually driven by precision ball screws and nuts, powered by direct-current servomotors.

Parts of the CNC Chucking and Turning Centers

The main parts of the CNC chucking and turning centers are the bed, headstock, cross-slide, carriage, turret, tailstock, servomotors, ball screws, hydraulic and lubrication systems, and the machine control unit (MCU).

Bed: The bed is usually made of high-quality cast iron, which is well suited to absorb the shock created by heavy cuts. It is usually a slant bed design (Fig. 10-12) of from 30° to 45°, which provides

Fig. 10-11   The CNC center-drive lathe is designed to machine parts from both sides at the same time. (*Cincinnati Milacron, Inc.*)

Fig. 10-12 The slant bed design provides for easy operator access and chip removal. (*Cincinnati Milacron, Inc.*)

easy access for the operator in loading and unloading of the parts. It also allows the chips and coolant to fall away from the cutting area to the bottom of the bed. Parallel surfaces are machined into the front of the cast bed, providing mounting tracks for the hardened bed ways.

Headstock: The headstock of the CNC chucking and turning centers (Fig. 10-13) transmit the maximum horsepower and torque from the motor to the spindle. These machines are available with a variety of motor sizes ranging from 5 to 75 horsepower and spindle speeds from 32 to 5500 revolutions per minute (r/min). The spindle speed is usually programmable in 1-r/min increments.

Tailstock: CNC chucking and turning centers can be equipped with different types of tailstocks—a manual tailstock similar to a standard engine lathe (Fig. 10-14), an automatic tape-controlled tailstock, or a tape-controlled swing-up tailstock. The tailstock travels on its own hardened and ground bearing ways. This allows the carriage to move past the tailstock when a short shaft is being held. It also eliminates the need to extend the quill of the tailstock to its maximum distance, thus maintaining greater rigidity of the part.

The *automatic tape-controlled tailstock* (Fig. 10-15) can be moved by tape command or manually by the operator using the switches on

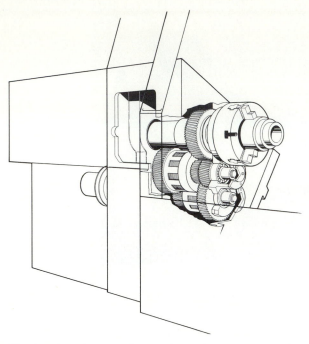

Fig. 10-13   The headstock transmits maximum horsepower and torque to the spindle. (*Cincinnati Milacron, Inc.*)

the MCU. Positioning, clamping, and unclamping of the tailstock to the bearing ways are done by hydraulic pressure. The tailstock is protected against collision with the indexing tools by a contact sensor which immediately stops the indexing motion on contact.

The *swing-up tailstock* (Fig. 10-16) adds flexibility and versatility to either the chucking or the turning center. It can swing up to support workpieces for external machining, and then swing away to allow the machine to perform internal work such as deep-hole drilling and boring.

Turrets: The type, style, and number of turrets on a CNC chucking or turning center will vary according to the size of the machine and individual manufacturer's specification. The more common types used are the drum turret, disk turret, and the square multitool turret. The turret is constructed so that it can hold eight or more inner and outer turning tools (Fig. 10-17). The indexing mechanism for these tools is capable of bidirectional indexing and a rapid traverse rate of approximately 400 in./min (100 m/min), which minimizes the noncutting time.

Fig. 10-14 A standard manual tailstock used to support the end of a workpiece. (*Cincinnati Milacron, Inc.*)

Fig. 10-15 The programmable tailstock can be moved manually or by tape or computer command. (*Cincinnati Milacron, Inc.*)

Fig. 10-16   The swing-up tailstock provides flexibility and versatility to the machine. (*Cincinnati Milacron, Inc.*)

Servo system: The servo system, which consists of servo drive motors, ball screws, and rotary resolvers, provides the fast, accurate movements and positioning of the X and Z axes slides. The rotary resolvers provide the system with unidirectional slide-positioning repeatability of ±0.0002 in. (0.005 mm). The rapid movement of the slides is approximately 200 to 400 in./min (5000 to 10000 mm/ min), and the feed rates are programmable in 0.100-in./min (2.54-mm/min) increments.

Fig. 10-17   A disk-type turret holding both inner and outer turning tools. (*Cincinnati Milacron, Inc.*)

Fig. 10-18   The MCU is used to control the machine for manual and automatic operation. (*Cincinnati Milacron, Inc.*)

MCU: The MCU (Fig. 10-18) allows the operator to program a part, edit a program, graphically display programs, store programs into memory, output programs to a punch or data line, perform comprehensive diagnostics, run a program manually or automatically, and perform many more functions and operations. These controls are designed and manufactured using the latest "state-of-the-art" technologies and features and will vary accordingly to individual manufacturer's requirements and specifications.

WORKHOLDING DEVICES

The demand for heavier metal removal rates and higher spindle speeds has created a need for high-performance chucks. Along with these performance requirements is the need for better and more secure gripping of the workpiece, quick-change jaws, chucks that can handle different-size workpieces, etc.

The most common workholding device on turning and chucking centers is a chuck. There is a wide variety of chucks, such as self-centering, counter-centrifugal, and collet, to suit various workpieces and machining conditions.

Self-Centering Chuck

Self-centering chucks are designed to move all jaws equally and simultaneously to center the part in the chuck. Self-centering chucks (Fig. 10-19) normally have higher gripping forces and are more accurate than other chuck types. These chucks are recommended for bar stock, forgings, castings, or turned parts which are located from the gripping diameter. The self-centering chuck is front-actuated and is arranged to hold collet pads in the master jaws for bar stock operations.

Countercentrifugal Chuck

The *countercentrifugal chuck* is one way manufacturers have met the need to better grip the workpiece at high speeds. The countercentrifugal chuck reduces the centrifugal force developed by the high r/min: counterweights pivot so that the centrifugal force tends to increase the gripping pressure, thus

Fig. 10-19 Self-centering chucks are recommended for bar stock, forgings, and castings. (*Cincinnati Milacron, Inc.*)

Fig. 10-20 The collet chuck provides precision holding for the workpiece. (*Hitachi Seiko Co., Ltd.*)

offsetting the outward forces developed by centrifugal force of the chuck jaws.

One of the disadvantages of these chucks is the tendency to increase the gripping pressure as the turning center slows down, which may damage the workpiece. An alternative method is to use elements of the chuck to lock the chuck jaws mechanically in their original position.

Countercentrifugal chucks come in a variety of sizes from 8 to 18 in. (200 to 450 mm) in diameter and can operate at spindle speeds of 5500 r/min for an 8-in. (200-mm) diameter chuck and 3500 r/min for a 12-in. (300-mm) diameter chuck. The repeatability of these chucks is 0.001 in. (0.02 mm).

Collet Chucks

The *collet chuck* (Fig. 10-20) is ideal for holding square, hexagonal, and round bar stock. The collet assembly consists of a drawtube, a hollow cylinder with master collets, and collet pads. Master collets are available with three or four gripping fingers and are referred to as either three-split or four-split design. The four-split design has better gripping power, but is less accurate. The

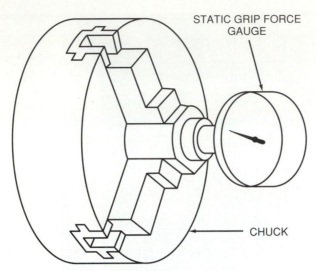

STATIC GRIP FORCE GAUGE

CHUCK

Fig. 10-21 Static gripping force determines the force per jaw being exerted on the workpiece when stopped. (*Cincinnati Milacron, Inc.*)

collet chucks are front-actuated, and the collet pads are sized for the diameter to be machined.

Chucks for Clamping Force

Two devices can be used to measure the clamping force that is put on the workpiece. One measures the static gripping force (Fig. 10-21), and one measures the dynamic gripping force. The *static gripping force* is the force per jaw exerted by the chuck on the workpiece when the spindle is stopped. The *dynamic gripping force* is the force per jaw exerted by the chuck on the workpiece when the spindle is running. Each individual chuck will have a specific clamping force.

Chuck Jaw Pressure Limitations

When the chuck jaw clamping pressure is set, the pressure must not exceed the maximum pressure stamped on the warning plate. If a greater pressure is used, high stress forces are created within the chuck, resulting in possible damage to the chuck or to the machine, which may cause personal injury. Typically, front-actuated chucks can be operated between 200 and 500 psi. Operating the chuck below 200 psi will cause insufficient clamping force on the part.

Always use the maximum chuck clamping pressure unless the pressure applied will damage the part.

Centrifugal Force and Speed Limitations

Centrifugal force imposes speed limitations upon all types of chucks. Centrifugal force, which increases as the speed of rotation increases, tends to throw the chuck jaws outward. This decreases the amount of the clamping force on the part. No chuck is immune from high internal stresses caused by centrifugal force; therefore, all chucks have a maximum rotation speed. This speed must not be exceeded under any circumstance.

Operating chucks at spindle speeds which are higher than their rated maximum speed will result in higher internal stresses and the loss of clamping force on the part. High stresses can cause jaw breakage or chuck component breakage, resulting in the release of the jaws and the workpiece.

Centrifugal force, in addition to increasing with speed, increases as the jaws are moved outward from the center line of rotation and increases as the jaws are made heavier. Do not mount top jaws so that they extend beyond the diameter of the chuck. Also, reduce the spindle speed when using special top jaw tooling configurations.

Changing Chuck Jaws

When it is necessary to frequently change the chuck jaws to accommodate different-size workpieces, or when additional machining operations require a different method of holding the workpiece, a quick jaw-changing system can be used. This quick jaw-changing system (Fig. 10-22A) can reduce the changing time from the usual 30 min to 1½ min or less. This optional feature can pay for itself in a very short period of time: it results in less machine downtime, and therefore the productivity of the machine improves dramatically.

Turning centers can also be equipped with an automatic jaw- or chuck-changing system (Fig. 10-22B). The more chuck-changing that is required, the more important the system becomes in providing less downtime of the turning center. When the automated jaw-changing system is called up by the NC program, it moves into position in front of the chuck. Jaws that are mounted in the chuck are removed and returned to their position in the magazine, and those required for the next operation are mounted into the chuck.

Some automated systems change one jaw at a time, while other systems are capable of changing all three jaws at once. Manually changing the chuck jaws could take the operator 20 to 30 min, but with new quick-change designs, the jaws or the inserts can be changed in 1 min or less.

Steadyrest/Follower Rest

When long, thin workpieces or shafts are machined, chatter usually occurs. The chatter or vibration can be minimized by means of a steadyrest (Fig. 10-23) or a follower rest (Fig. 10-24).

WORKPIECE-HANDLING SYSTEM

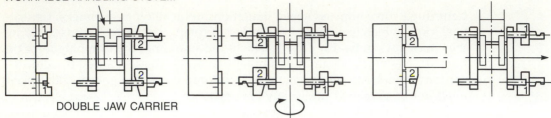

DOUBLE JAW CARRIER

Fig. 10-22A A quick jaw-changing system can reduce the usual time of changing jaws from 30 min to 1½ min or less. (*Rohm Products of America*)

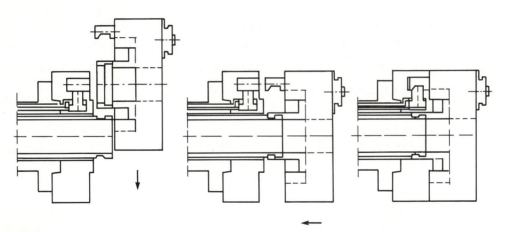

Fig. 10-22B An automatic jaw- or chuck-changing system reduces the amount of machine downtime and increases productivity. (*Forkardt, Inc.*)

The steadyrest is mounted to the bed of the turning center and can be programmed to open or close automatically, providing *feed-through* capabilities for the workpiece. The steadyrest uses constant hydraulic pressure applied to the support rollers, enabling it to adjust to the proper workpiece diameter. The support provided by the steadyrest reduces chatter in the workpiece and allows the machine to operate at higher and more efficient speeds and feeds.

When a four-axis chucking or turning center is used, a follower rest can be mounted in the lower turret. The turret can be programmed to move at the desired rate of travel, providing constant support to the workpiece, while the

Fig. 10-23   Feed through steadyrest adjusts to various diameters and increases stability.

cutting tool mounted in the upper turret does the machining. Used in this manner, the follower rest could also perform the functions of a steadyrest: by arbitrarily positioning the lower turret, the follower rest would support the workpiece at a desired location (Fig. 10-25).

If there is insufficient workpiece chucking area, the steadyrest or follower rest can be used as a safety measure to support the workpiece, preventing it from being thrown out of the machine during the cutting operation.

Fig. 10-24   Two follower rests being used to support a long thin shaft. (*Cincinnati Milacron, Inc.*)

Fig. 10-25   A follower rest, held in the lower turret, is automatically positioned to provide support for the workpiece. (*Yamazaki Machinery Works, Ltd.*)

TOOLING SYSTEMS AND CUTTING TOOLS

Tooling systems for turning and chucking centers may vary with individual manufacturer's specification. It is important to remember that the success of any turning operation will depend on the accuracy of the tooling system and the cutting tools being used. A typical tooling system (Fig. 10-26) consists of toolholders, boring bar holders, facing and turning holders, and drill sockets.

Cutting Tools

There is a variety of types and styles of indexable insert tooling available for chucking and turning centers (Fig. 10-27). This type of tooling allows the operator to change the indexed inserts at the machine tool instead of removing the cutting tool for resharpening and replacing it with another tool.

Carbide Inserts

Because of the tremendous variety of carbide inserts available (Fig. 10-28), it is important that the correct grade and shape be selected for the type of material and machining application required. Two main considerations in selecting the proper insert are:

1. Is the insert capable of cutting the required contours?
2. Does it have sufficient strength to complete the cut?

TOOLING SYSTEM

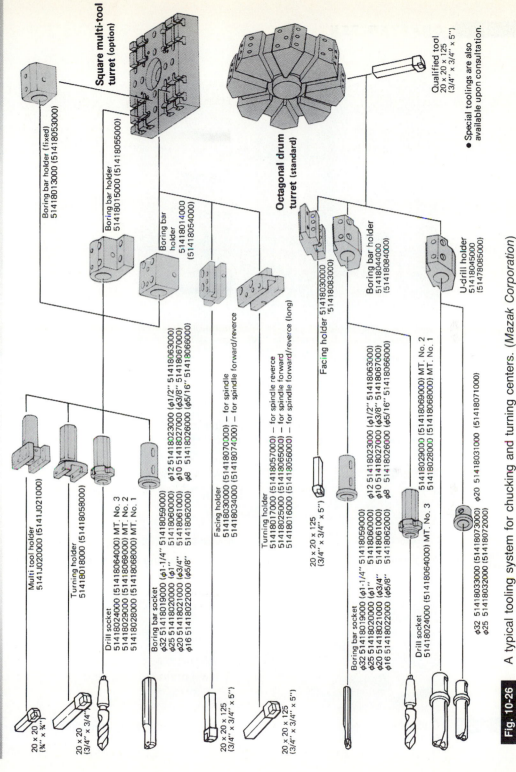

Fig. 10-26 A typical tooling system for chucking and turning centers. *(Mazak Corporation)*

Boring bar holder (fixed) 51418013000 (51418053000)

Boring bar holder 51418015000 (51418055000)

Boring bar holder 51418014000 (51418054000)

Square multi-tool turret (option)

Octagonal drum turret (standard)

Qualified tool 20 x 20 x 125 (3/4" x 3/4" x 5")

• Special toolings are also available upon consultation.

Boring bar holder 51418044000 (51418084000)

Facing holder 51418030000 (51418083000)

U-drill holder 51418045000 (51478085000)

Multi tool holder 5141J020000 (5141J0210000)

Turning holder 51418018000 (51418058000)

Drill socket
51418024000 (51418064000) MT. No. 3
51418029000 (51418069000) MT. No. 2
51418028000 (51418068000) MT. No. 1

Boring bar socket
φ32 51418019000 (φ1-1/4" 51418059000)
φ25 51418020000 (φ1" 51418060000)
φ20 51418021000 (φ3/4" 51418061000)
φ16 51418022000 (φ5/8" 51418062000)

φ12 51418023000 (φ1/2" 51418063000)
φ10 51418027000 (φ3/8" 51418067000)
φ8 51418026000 (φ5/16" 51418066000)

Facing holder
51418030000 (51418070000) — for spindle
51418034000 (51418074000) — for spindle forward/reverce

Turning holder
51418017000 (51418057000) — for spindle reverce
51418025000 (51418065000) — for spindle forward
51418016000 (51418056000) — for spindle forward/reverce (long)

20 x 20 x 125
(3/4" x 3/4" x 5")

20 x 20 x 125
(3/4" x 3/4" x 5")

Boring bar socket
φ32 51418019000 (φ1-1/4" 51418059000)
φ25 51418020000 (φ1" 51418060000)
φ20 51418021000 (φ3/4" 51418061000)
φ16 51418022000 (φ5/8" 51418062000)

φ12 51418023000 (φ1/2" 51418063000)
φ10 51418027000 (φ3/8" 51418067000)
φ8 51418026000 (φ5/16" 51418066000)

Drill socket
51418024000 (51418064000) MT. No. 3
51418029000 (51418069000) MT. No. 2
51418028000 (51418068000) MT. No. 1

φ32 51418033000 (51418073000) φ20 51418031000 (51418071000)
φ25 51418032000 (51418072000)

20 x 20
(3/4" x 3/4")

20 x 20
(3/4" x 3/4")

20 x 20 x 125
(3/4" x 3/4" x 5")

20 x 20 x 125
(3/4" x 3/4" x 5")

Fig. 10-27   A variety of cutting tools used on chucking and turning centers. (*Kennametal Inc.*)

Fig. 10-28   Carbide-insert cutting tools help to increase productivity and reduce machining time. (*Carboloy Inc.*)

There is a complex relationship between the cutting condition, speed and feed, and the tool life. Some of the factors that will affect tool life are:

- Workpiece material
- Insert grade
- Insert geometry
- Insert nose radius
- Depth of cut
- Coolant

Tool Monitoring

When machining is done on a turning center, tool wear and breakage require the continuous attention of the operator. A tool-monitoring system (Fig. 10-29) can be substituted for the skilled operator's eyes and ears, signaling in a variety of ways the need to replace tools that are worn and broken.

There are many types of tool-monitoring systems available, and the way they detect tool wear varies with the manufacturer. Given that a worn or dull tool requires more power to machine a workpiece than a sharp tool, the most common method of determining wear is from the power or force it takes to drive the cutting tool. The tool-monitoring system measures the load on the main spindle drive motor. This is done in two stages:

1. When the machine is set up and ready to produce a workpiece, the normal machining cycle is run with new tools and all speeds and feeds at 100 percent of programmed rate.

2. When the workpiece is completed, a second machining cycle is run, this time without contacting the workpiece.

Fig. 10-29 A tool-monitoring system can inform the operator when a cutting tool is worn and needs replacing. (*Cincinnati Milacron, Inc.*)

From these two cycles, the monitoring system can calculate the net machining forces and torque for every portion of the part program where monitoring is desired. Once the limits have been set, the monitoring system will signal the operator when machining forces and torque exceed acceptable limits and, in some cases, automatically reduce the speed or feed to compensate for the dullness of the cutting tool.

Some other methods of detecting tool wear and breakage are by:

- Sound
- Vibration
- Measuring heat
- Electrical resistance
- Radioactivity
- Optical magnification

Regardless of the system of detection used, the tool-wear-monitoring system provides such benefits as:

- Broken tool detection
- Machine protection
- Improved productivity
- Reduced operator attention
- Worn tool detection

In-Process Gaging

An in-process gaging system (Fig. 10-30) is a way of monitoring what is happening to the workpiece and tools during the machining operations. It can also be used to compensate for tool wear and thermal growth, determine tool offsets and locate workpieces, and in datuming, inspection, etc.

The probe works basically by sending a signal to the machine control as soon as the probe has been deflected in any direction upon making contact with the workpiece or the tool. In-cycle gages are omnidirectional, which means that the sensing probes will sense any $\pm X$, $\pm Y$, or $+Z$ direction. Once the probe stylus has been deflected, a signal is automatically sent to the control where the data can be acted on.

On chucking and turning centers, the probes can be mounted either in a toolholder or in a turret. The probes can be selected the same way a cutting tool is selected—by calling it up in the machining program—to check a tool for wear and make the appropriate compensation, or to check a part for size between machining operations.

The use of in-process gaging helps to reduce operator errors in setup and allows for inspection of fully machined parts in the machine.

Fig. 10-30 In-process gaging measures the workpiece and can compensate for tool wear during a machining operation. (*Cincinnati Milacron, Inc.*)

SPEEDS AND FEEDS

There are charts available as guidelines for setting surface speed, feed, and depth of cut (Fig. 10-31). When choosing a surface speed, look at the specific conditions that relate to the workpiece being machined. Interrupted cuts, large depth of cuts, long continuous cuts, surface scale, high feed rates, no coolant, rigidity of setup or workpiece—all would indicate a reduction in the recommended surface speed, whereas uninterrupted cuts, light feed and depth of cuts, short length of cuts, smooth prefinished materials, flood coolant, and rigid setup would allow for recommended surface speeds and still maintain an acceptable tool life.

The correct feed should be maintained for all tools. Incorrect feed rates produce a number of problems. Too slow a feed rate can cause problems with chip control and reduce cutter life. Too fast a feed rate can cause insert chipping and breakage and reduced cutter life.

| Material | Finish* | Depth of cut | | High-speed steel tool | | | Carbide tool | | | |
| | | | | Speed sfpm (CS) | Feed | | Speed sfpm (CS) | | Feed | |
		in.	mm		ipr	mm	Brazed	Throw-away	ipr	mm
Aluminium alloys	R	0.150	3.81	600	0.015	0.38	1100	1500	0.020	0.50
Wrought	F	0.025	0.63	800	0.007	0.17	1400	1800	0.010	0.25
Brass	R	0.150	3.81	400	0.015	0.38	800	925	0.020	0.50
330-340-353	F	0.025	0.63	480	0.007	0.17	960	1100	0.007	0.17
Cast iron	R	0.150	3.81	145	0.015	0.38	500	550	0.020	0.50
Soft	F	0.025	0.63	185	0.007	0.17	650	725	0.010	0.25
Cast iron	R	0.150	3.81	80	0.015	0.38	300	340	0.015	0.38
Hard	F	0.025	0.63	120	0.007	0.17	360	410	0.007	0.17
Carbon steel (Lo)	R	0.150	3.81	120	0.015	0.38	400	485	0.020	0.50
1010-1020	F	0.025	0.63	160	0.007	0.17	475	625	0.007	0.17
Carbon steel (Med)	R	0.150	3.81	75	0.015	0.38	300	375	0.020	0.50
1030-1055	F	0.025	0.63	105	0.007	0.17	385	475	0.007	0.17
Carbon steel (Hi)	R	0.150	3.81	65	0.015	0.38	275	345	0.015	0.38
1060-1095	F	0.025	0.63	85	0.007	0.17	360	440	0.007	0.17
Alloy steel (Med C)	R	0.150	3.81	90	0.015	0.38	300	400	0.020	0.50
4130-4140	F	0.025	0.63	120	0.007	0.17	400	500	0.007	0.17
Tool steel (HS)	R	0.150	3.81	60	0.015	0.38	250	290	0.015	0.38
M-3, M-4, M-7	F	0.025	0.63	65	0.007	0.17	275	320	0.007	0.17
Stainless steel	R	0.150	3.81	105	0.015	0.38	425	475	0.015	0.38
300 Series	F	0.025	0.63	125	0.007	0.17	475	520	0.007	0.17
Stainless steel	R	0.150	3.81	150	0.015	0.38	475	525	0.015	0.38
400 Series	F	0.025	0.63	170	0.007	0.17	525	590	0.007	0.17

*R designates rough cut depths; F designates finish cut depths.

Note: These cutting speeds can be used to calculate spindle speeds; however, such speeds are approximate and should be adjusted for type and condition of machine, exact dimensions of operation, and type of material being machined.

Fig. 10-31   Recommended cutting speed and feeds for high-speed steel and cemented carbide cutting tools.

MATERIAL HANDLING SYSTEMS

There are a number of options that can be added to the CNC chucking and turning centers to enhance their performance and productivity. Some of these options are the bar feeder, parts catcher, parts loader/unloader, chip conveyor, and robot loader.

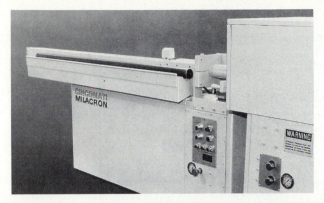

Fig. 10-32 The bar feeder can handle long lengths of bars to eliminate loading of individual part blanks. (*Cincinnati Milacron, Inc.*)

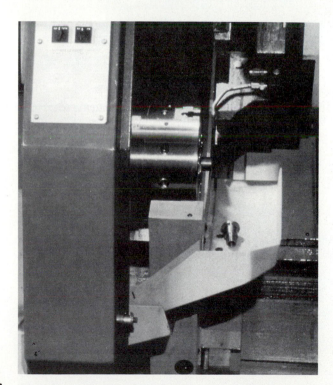

Fig. 10-33 The parts catcher deposits finished parts outside the machine. (*Cincinnati Milacron, Inc.*)

Fig. 10-34   The part loader/unloader places individual part blanks into the machine and removes the finished part. (*Cincinnati Milacron, Inc.*)

Fig. 10-35   The chip conveyor removes the chips from the turning center and places them into storage bins. (*Cincinnati Milacron, Inc.*)

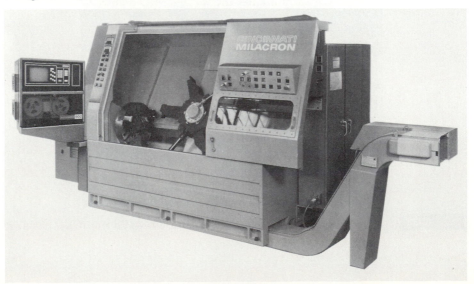

Fig. 10-36 Robot loaders are used for automated workhandling for turning centers. (*Cincinnati Milacron, Inc.*)

Bar Feeder

The bar feeder (Fig. 10-32) is capable of handling 6-ft (2-m) and 12-ft (4-m) bar lengths. This eliminates the loading of individual part blanks.

Parts Catcher

The parts catcher (Fig. 10-33) complements the bar feeder and deposits the machined parts outside of the machine. When the bar feeder and parts catcher are used, the loading and unloading time is reduced.

Parts Loader/Unloader

The parts loader/unloader (Fig. 10-34) allows individual part blanks of 0.75 to 2.0 in. (20 to 50 mm) in diameter and 1.0 to 2.0 in. (25 to 50 mm) in length to be loaded and unloaded in approximately 6 s.

Chip Conveyor

The chip conveyor (Fig. 10-35) picks up all the chips from the bed of the machine. The chips produced by the cutting cycle fall freely onto the chip conveyor track because of the slant bed design of the machine. The chips are then transported by the conveyor system out of the bottom of the machine into containers for storage and recycling.

Robot Loaders

Robot loaders (Fig. 10-36) communicate with the MCU and are capable of doing a variety of different operations, such as loading and unloading parts, storing and retrieving parts from pallets, transporting parts to gaging stations, and changing chuck jaws. Dedicated robot loaders represent the major trend in automated workhandling for turning centers.

FUNCTION CODES

Many preparatory and miscellaneous function codes are the same for turning centers, machining centers, and wire-cut electrical discharge machines (EDMs). However, because of the nature of work produced on each type of machine, certain differences exist in the coding systems to suit each machine. It is very important that the programmer recognize that differences do exist in order to properly program each machine. The use of an incorrect code might result in scrapped work, damage to the machine or cutting tool, or no response from the MCU. The codes used on turning and machining centers are listed as follows:

Preparatory Codes for Turning Centers

G00	Rapid positioning
G01	Linear interpolation
G02	Circular interpolation counterclockwise (CCW)
G03	Circular interpolation clockwise (CW)
G04	Dwell
G20	Inch data input
G21	Metric data input
G27	Zero return check
G28	Zero return
G29	Return from zero
G32	Thread cutting
G40	Tool tip R compensation cancel
G41	Tool tip R compensation right
G42	Tool tip R compensation left
G50	Absolute coordinate preset

G52	Local coordinates preset
G54–G59	Work coordinates system selection
G70	Finish cycle
G71	OD, ID rough cutting cycle
G72	End surface rough cutting cycle
G73	Closed-loop cutting cycle
G74	End surface cutting off cycle
G75	OD, ID cutting off cycle
G76	Thread-cutting cycle
G90	Cutting cycle A
G92	Thread-cutting cycle
G94	Cutting cycle B
G96	Constant surface speed
G97	Constant surface speed cancel
G98	Feed per time
G99	Feed per spindle rotation

Miscellaneous Functions for Turning Centers

M00	Program stop
M01	Optional stop
M02	End of program
M03	Spindle rotation—normal
M04	Spindle rotation—reverse
M05	Spindle stop
M08	Coolant on
M09	Coolant off
M10	Chuck—clamping
M11	Chuck—unclamping
M12	Tailstock spindle out
M13	Tailstock spindle in
M17	Toolpost rotation normal
M18	Toolpost rotation reverse
M21	Tailstock forward
M22	Tailstock backward
M23	Chamfering on

M24	Chamfering off
M30	Reset and rewind
M31	Chuck bypass on
M32	Chuck bypass off
M41	Spindle speed—low range
M42	Spindle speed—high range
M73	Parts catcher out
M74	Parts catcher in
M98	Call subprogram
M99	End subprogram

SAMPLE MACHINING PROGRAM

Turning centers manufactured by machine tool builders will vary: some mount the cross-slide on the slant bed, while others mount the cross-slide on a vertical support. Regardless of the construction of the machine, they all operate on the X (cross-slide movement) and Z (saddle movement) axes. The positive X ($+$X) moves the cutting tool toward the center line of the spindle, while the negative X ($-$X) moves the cutting tool away from the spindle center line. The positive Z ($+$Z) moves the saddle or cutting tool away from the headstock, while a negative Z ($-$Z) moves the saddle or cutting tool toward the headstock. Both incremental and absolute programming can be used on most turning centers.

The sample part shown in Fig. 10-37A will be used to program some common machining operations performed on a chucking turning center. Two passes will be made over the workpiece: one for rough turning and the other for finish turning operations.

The Machining Program

001

Program number

%

Rewind stop code

N010   G20

Inch input data

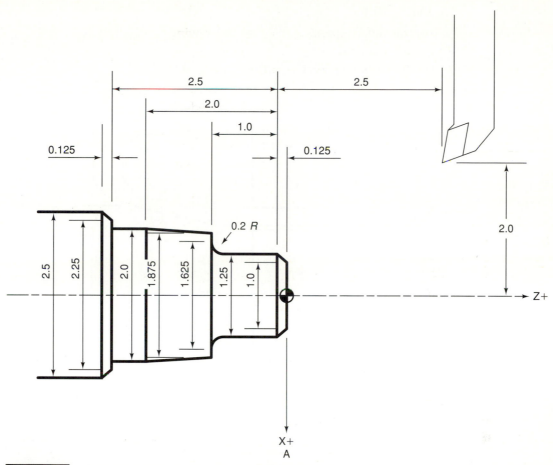

Fig. 10-37 A sample part which can be machined efficiently on a turning center. (Note: Parts B through F of this illustration appear on pages 282 to 286.)

Roughing Cut

N020   G50 X − 2.0 Z2.5 S1000

G50: absolute coordinate preset—the reference point for commanding the start point of individual cutting tools.

S1000: maximum spindle speed 1000 r/min.

N030   G00 T0101 M41

Tool number 01—activate offset number 01.

M41: low speed range (spindle).

N040 **G96 S100 M03**

G96: constant surface speed.

S100: speed 100 sf/min.

M03: spindle rotation on clockwise (CW).

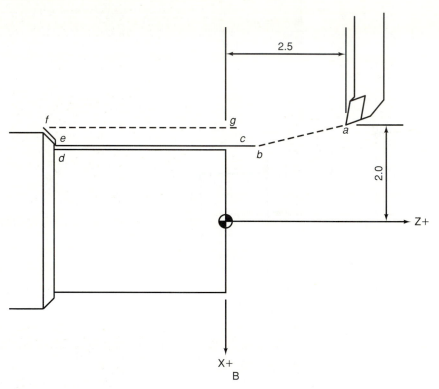

N050 **G00 X − 2.02 Z.1 M08**

G00: rapid move a–b.

M08: coolant on.

N060 **G01 Z.01 F.120**

Linear move b–c.

F120: feed rate in./r (0.120).

N070 **Z2.49 F.012**

Linear move c–d.

N080 **X − 2.27**

Linear move d–e.

N090   X – 2.52 Z – 2.625
Linear move e–f (chamfer).

N100   G00 Z.01
Rapid move f–g.

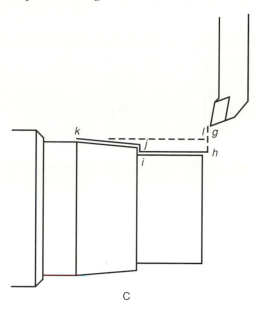

N110   X – 1.645
Rapid move g–h.

N120   G01 Z.99
Linear move h–i.

N130   X – 1.895
Linear move i–j.

N140   X – 2.02 Z – 2.0
Linear move j–k (taper).

N150   G00 Z.01
Rapid move k–l.

N160   X – 1.27
Rapid move l–m.

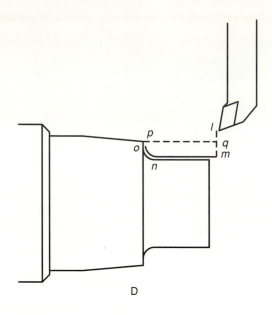

N170   G01 Z.79
Linear move m–n.

N180   G03 X − 1.645 Z.99 R.2
G03: circular interpolation clockwise (CW) n–o.
0.200 radius.

N190   G01 X − 1.895
Linear move o–p.

N200   G00 Z.01
Rapid move p–q.

N210   X − 1.02
Rapid move q–m.

N220   G01 X − 1.27 Z − .135
Linear move m–r (chamfer).

N230   G00 X − 4.0 Z2.5 M05
Rapid move r–a.
M05: spindle stop.

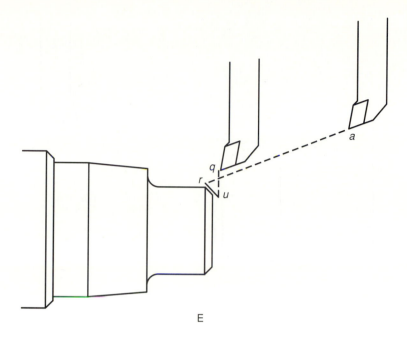

N240   T0100

Tool number 01—activate offset number 00 (00 canceled tool offset).

Finishing Cut

N250   G50 X − 2.0 Z2.5 S2000

G50: absolute coordinate preset for cutting tools.

S2000: maximum spindle speed 2000 r/min.

N260   T0202 M42

Tool number 02—activate offset number 02.

M42: high speed range (spindle).

N270   G96 S150 M03

G96: constant surface speed.

S150: speed 150 sft/min.

M03: spindle rotation on clockwise (CW).

N280   G00 X − 1.0 Z.2

Rapid move to start of finish cut a–t.

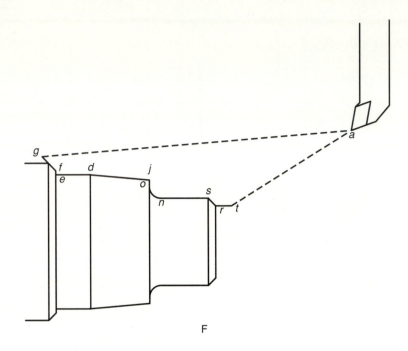

F

N290   G01 Z.1 F.040
Linear move t–r.
F.040: feed rate in./r (0.040).

N300   X – 1.25 Z – .125 F.006
Linear move r–s (chamfer).
F.006: feed rate.

N310   Z – .8
Linear move s–n.

N320   G03 X – 1.625 Z – 1.0 R.2
G03: circular interpolation clockwise (CW) n–o.
0.200 radius.

N330   G01 X – 1.875
Linear move o–j.

N340   X2.0 Z – 2.0
Linear move j–d (taper).

N350　　Z − 2.5
Linear move d–e.

N360　　X − 2.25
Linear move e–f.

N370　　X − 2.5 Z − 2.75 M09
Linear move f–g (chamfer).
M09: coolant off.

N380　　G00 X4.0 Z2.5 M05
Rapid move g–a.
M05: spindle stop.

N390　　T0200
Tool number 02—activate offset number 00 (00 canceled tool offset 02).

N400　　M30
End of program.

%
Rewind stop code.

REVIEW QUESTIONS

Chucking and Turning Centers

1. Define the term *retrofit*.

2. How did early NC lathes make contour cuts?

3. List five improvements found on today's CNC chucking and turning centers.

Types of Turning Centers

4. List five internal machining operations that can be performed on a chucking center.

5. How would long parts be supported and machined?

6. For what type of machining are turning centers mainly designed?

7. List four advantages of machines having dual turrets.

8. Briefly describe a CNC center-drive lathe.

Turning and Chucking Center Parts

9. What are the main parts of a CNC chucking and turning center?

10. Why is the bed usually made of high-quality cast iron?

11. Why is the bed of chucking and turning centers usually slanted?

12. What is the purpose of the headstock?

13. What is the usual programmable spindle speed?

14. Name and state the purpose of three different types of tailstocks.

15. Why does the tailstock travel on its own bedways?

16. How is the automatic tape-controlled tailstock moved?

17. How is the tailstock protected against collisions with the indexing tools?

18. Name the three most commonly used types of turrets.

19. What is the repeatability of the servo system, and how is this obtained?

20. List seven functions of the MCU.

Workholding Devices

21. What is the most common workholding device used?

22. List three different types of chucks used on chucking centers.

23. Name two advantages of self-centering chucks.

24. List the advantages and disadvantages of countercentrifugal chucks.

25. What is the repeatability of the countercentrifugal chuck?

26. What type of workpieces are best suited for machining with collet chucks?

27. Briefly define:
   (a) Static gripping force
   (b) Dynamic gripping force

28. Why is correct chuck clamping pressure important?

29. Why is it important not to operate chucks at speeds higher than their rated maximum speed?

30. Briefly describe what is meant by automatic chuck jaw-changing.

31. Name four advantages of using a follower rest or steadyrest when machining long, thin shafts.

Tooling Systems

32. What does a typical tooling system consist of?

33. What are the advantages of indexable insert tooling?

34. Name two main factors which should be considered when selecting carbide inserts.

35. List six factors that will affect tool life.

36. What cutting conditions would indicate a reduction in the cutting speed used?

37. What cutting conditions would indicate that recommended cutting speeds be used?

38. List two problems that result from too slow a feed or from too fast a feed rate.

Material Handling Devices

39. List and state the purpose of six material handling devices used with chucking and turning centers.

ELEVEN

Electrical Discharge Machining

Electrical discharge machining, commonly known as EDM, is a process that is used to remove metal through the action of an electrical discharge of short duration and high current density between the cutting tool and the workpiece (Fig. 11-1). This principle of removing metal by an electric spark has been known for quite some time. In 1889, Friedrich Paschen, the German physicist, explained the phenomenon and devised a formula that would predict its arcing ability in various materials. The EDM process can be compared with a miniature version of a lightning bolt striking a surface, creating a localized intense heat, and melting away the work surface.

After completing
this chapter,
you should be
able to:

1. Describe the principle of EDM

2. State the advantages and applications of the electrical discharge process

3. Program a wire-cut EDM to produce a variety of parts

EDM has proved especially valuable in the machining of super-tough, electrically conductive materials such as the new space-age alloys. These metals would be difficult to machine by conventional methods, but EDM has made it

Fig. 11-1 A controlled spark removes metal during electrical discharge machining. (*Cincinnati Milacron, Inc.*)

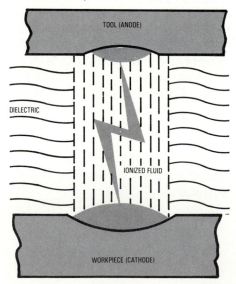

TOOL (ANODE)

DIELECTRIC

IONIZED FLUID

WORKPIECE (CATHODE)

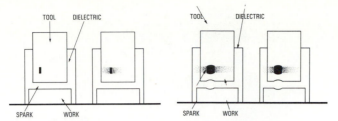

Fig. 11-2 The four stages of a single discharge. (*Cincinnati Milacron, Inc.*)

relatively simple to machine intricate shapes that would be impossible to produce with conventional cutting tools. New applications for this machining process are continually found in the metalworking industry. It is finding wide use in tool- and diemaking to accurately produce dies whose punch and die cavity have intricate shapes.

PRINCIPLE OF EDM

EDM is a controlled metal removal process whereby an electric spark is used to cut (erode) the workpiece, which then takes the shape opposite to that of the cutting tool or electrode. The electrode and the workpiece are both submerged in a *dielectric fluid*, which is generally a light lubricating oil. This dielectric fluid should be a nonconductor (or poor conductor) of electricity. A *servomechanism* maintains a gap of about 0.0005 to 0.001 in. (0.01 to 0.02 mm) between the electrode and the work, preventing them from coming into contact with each other. A direct current of low voltage and high amperage is delivered to the electrode at the rate of approximately 20,000 hertz (Hz). These electrical energy impulses become sparks which jump the gap between the electrode and the workpiece through the dielectric fluid (Fig. 11-2). Intense heat is created in the localized area of the spark impact; the metal melts and a small particle of molten metal is expelled from the workpiece. The dielectric fluid, which is constantly being circulated, carries away the eroded particles of metal and also helps in dissipating the heat caused by the spark.

There are two types of EDM machines used in industry: the *vertical* EDM machine and the *wire-cut* EDM machine (Fig. 11-3). Since the wire-cut EDM is generally used for machining complex forms which require NC programming, only this type will be discussed in detail.

A

Fig. 11-3 Electrical discharge machines. (A) Vertical EDM is used for sinking cavities in work-pieces. (*Continued on page 295*)

WIRE-CUT EDM

The wire-cut EDM is a discharge machine which uses NC movement to produce the desired contour or shape on a part. It does not require a special-shaped electrode; instead, it uses a continuous traveling wire under tension as the electrode. The electrode or cutting wire can be made of brass, copper, or any other electrically conductive material ranging in diameter from 0.002 to 0.012 in. (0.05 to 0.30 mm). The path that the wire follows is controlled along a two-axis (XY) contour, eroding (cutting) a narrow slot through the work-piece. This controlled movement is continuous and simultaneous in increments of 0.00005 in. (0.001 mm). Any contour may be cut to a high degree of accuracy and is repeatable for any number of successive parts. A *dielectric*

B

Fig. 11-3 (*Continued*) (B) wire-cut EDM is used for machining complex forms. (*LeBlond-Makino Machine Tool Company*)

fluid, usually deionized water which is constantly being circulated, carries away the eroded particles of metal. The dielectric fluid maintains the proper conductivity between the wire and the workpiece and assists in reducing the heat caused by the spark.

Parts of the Wire-Cut EDM

The main parts of the wire-cut EDM are the bed, saddle, table, column, arm, UV axis head, wire feed and dielectric systems, and machine control unit (MCU) (Fig. 11-4).

Bed: The bed is a heavy, rugged casting used to support the working parts of the wire-cut EDM. Guide rails ("ways") are machined on the top section, and these guide and align major parts of the machine.

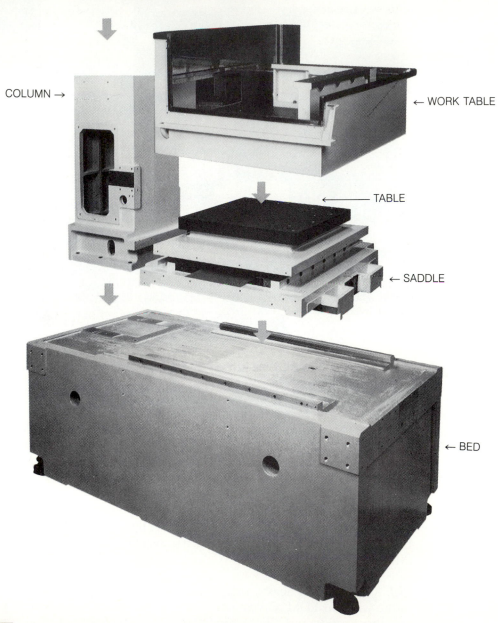

← BEAM

COLUMN →

← WORK TABLE

← TABLE

← SADDLE

← BED

Fig. 11-4 The main parts of a wire-cut EDM. (*LeBlond-Makino Machine Tool Company*)

Saddle: The saddle is fitted on top of the guide rail and may be moved in the XY direction by feed servomotors and ball lead screws.

Table: The table, mounted on top of the saddle, is U-shaped and contains a series of drilled and tapped holes equally spaced around the top surface. These are used for the workpiece holding and clamping devices.

Column: The column, which is attached to the bed, supports the wire feed system, the UV axes, and the capacitor switches.

Wire Feed System: The wire feed system is used to provide a continual feed of new wire (electrode) for the cutting operation. The wire is fed from a supply spool through a series of guides and guide rollers which apply tension to the wire. The wire travels in a continuous path past the workpiece, and the used wire is rewound on a takeup spool.

Dielectric System: The dielectric system contains filters, an ion exchanger, and a cooler. This system provides a continuous flow of clean deionized water at a constant temperature. The deionized water stabilizes the cutting operation, flushes away particles of electrode and workpiece material that have been eroded, and cools the workpiece.

MCU: The MCU can be separated as three individual panels.

1. The control panel for setting the cutting conditions
2. The control panel for machine setup
3. The control panel for manual data input (MDI) and cathode-ray tube (CRT) character display (Fig. 11-5)

MACHINE OPERATING SYSTEMS

Various systems on wire-cut EDM machines play an important part in the efficient operation of the machine tool. Servo systems, dielectric fluid, electrode, and the MCU are the main operating components of wire-cut EDM machines.

The Servomechanism

The EDM power supply controls the cutting current levels and the feed rate of the drive motors. It also controls the travelling speed of the wire (Fig. 11-6).

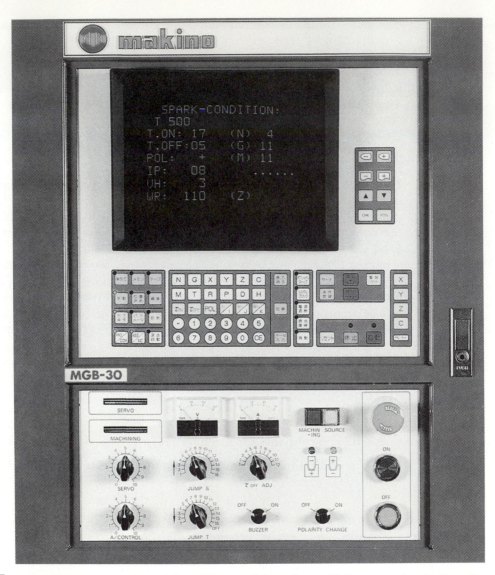

Fig. 11-5 The MCU is used to control the cutting conditions along with the manual and automatic operation of the machine. (*LeBlond-Makino Machine Tool Company*)

EDM machines are equipped with a servo control mechanism that automatically maintains a constant gap of approximately 0.001 to 0.002 in. (0.02 to 0.05 mm) between the wire and the workpiece. It is important that there be no physical contact between the wire (electrode) and the workpiece; otherwise arcing could damage the workpiece and break the wire. The servomech-

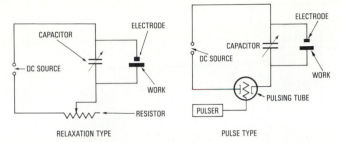

RELAXATION TYPE   PULSE TYPE

Fig. 11-6 Electrical discharge power supply circuits control the traveling speed of the wire. (*Cincinnati Milacron, Inc.*)

anism also advances the wire into the workpiece as the operation progresses and senses the work-wire spacing and slows or speeds up the drive motors as required to maintain the proper arc gap. Precise control of the gap is essential to a successful machining operation. If the gap is too large, ionization of the dielectric fluid does not occur and machining cannot take place. If the gap is too small, the wire will touch the workpiece, causing it to melt and break.

Precise gap control is accomplished by a circuit in the power supply comparing the average gap voltage to a preselected reference voltage. The difference between the two voltages is the input signal, which tells the servomechanism how far and how fast to feed the wire and when to retract from the workpiece. This is usually indicated by a reverse servo light.

When chips in the spark gap reduce the voltage below a critical level, the servomechanism causes the wire to retract until the chips are flushed out by the dielectric fluid. The servo system should not be too sensitive to "short-lived" voltages caused by chips being flushed out; otherwise the wire would be constantly retracting, thereby seriously affecting machining rates.

The Dielectric Fluid

One of the most important factors in a successful EDM operation is the removal of the particles (chips) from the working gap. Flushing these particles out of the gap with the dielectric fluid will produce good cutting conditions, while poor flushing will cause erratic cutting and poor machining conditions.

The dielectric fluid in the wire-cut EDM process is usually deionized water. This is tap water that is circulated through an ion-exchange resin. The deionized water makes a good insulator, while untreated water is a conductor and is not suitable for the EDM process. The ion exchanger is a compound of positive ion exchange resin (cation) and negative ion exchange resin (anion). When water is applied to this compound, the ion exchange reaction starts and is continuously repeated until all the impurities are completely removed from the water, thus producing pure water.

The amount of deionization is measured by its specific resistance. For most

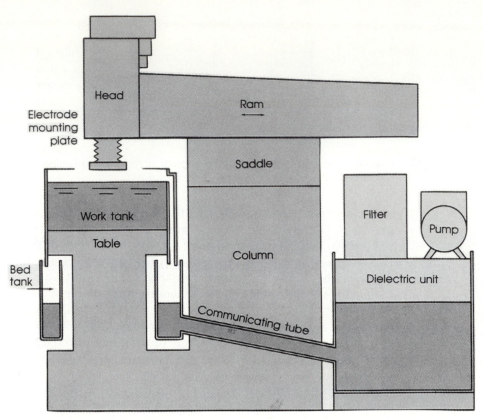

Fig. 11-7 The dielectric fluid system maintains conductivity and assists in reducing heat caused by the spark. (*LeBlond-Makino Machine Tool Company*)

operations, the lower the resistance, the faster the cutting speed. However, the resistance of the dielectric fluid should be much higher when carbides and high-density graphites are cut (Fig. 11-7).

The dielectric fluid used in the wire-cut EDM process serves several functions:

1. It helps to initiate the spark between the wire (electrode) and the workpiece.

2. It serves as an insulator between the wire and the workpiece.

3. It flushes away the particles of disintegrated wire and workpiece to prevent shorting.

4. It acts as a coolant for both the wire and the workpiece.

The dielectric fluid must be circulated under constant pressure if it is to flush away the particles and assist in the machining process. The flow of water is governed by two control valves: one controlling the water flowing above the workpiece and one controlling the flow under the workpiece. When starting to supply the water for the cutting process, apply a steady amount "downstream" and then slowly apply the water from underneath until there is a flaring effect at the top surface of the workpiece (Fig. 11-8).

If red sparks occur during the cutting operation, the water supply is inadequate. To overcome this problem, increase the flow of water until blue sparks appear. *Note:* Excessive water flow can cause the wire to be deflected, causing erratic cutting and loss of machining accuracy.

The Electrode

The electrode in wire-cut EDM is a spool of wire ranging from 0.002 to 0.012 in. (0.05 to 0.30 mm) in diameter and from 2 to 100 lb (0.90 to 45.36 kg) in weight. The length of wire on a spool can provide over 500 h of unattended machining time. The electrode of the wire-cut machine continuously travels from a supply spool to a takeup spool so that it is constantly renewed. When this type of electrode is used, the wear on the wire does not affect the accuracy of the cut because new wire is being fed past the workpiece continuously at rates from a fraction of an inch to several inches per minute. Both the

Fig. 11-8   A steady stream of dielectric must be supplied to the workpiece for the cutting operation. (*LeBlond-Makino Machine Tool Company*)

electrode wear and the material removal rate from the workpiece depend on such things as the material's thermal conductivity, its melting point, and the duration and intensity of the electrical pulses. As in conventional machining, some EDM materials have better cutting and wearing qualities than others; therefore, electrode materials must have the following characteristics:

1. Be a good conductor of electricity
2. Have a high melting point
3. Have a high tensile strength
4. Have good thermal conductivity
5. Produce efficient metal removal from the workpiece

Research and experimentation are continually increasing metal removal rates and finding good, economical materials for the manufacture of EDM wire electrodes. Brass, copper, tungsten, molybdenum, and zinc are some of the materials which have found certain applications as electrode materials. The most widely used material is brass wire 0.008 in. (0.20 mm) in diameter. A normal overcut of about 0.001 in. (0.02 mm) per side will produce an internal corner with a radius of about 0.005 in. (0.12 mm). Smaller diameters of brass wire are also used for many applications. Tungsten and molybdenum wire, which have a very high melting point and high tensile strength, permit small-diameter wire of 0.002 in. (0.05 mm) in diameter to be used for cutting fine radii and intricate shapes.

Stratified wire, which consists of a copper core with a thin surface layer of zinc, gives the wire the high conductivity of copper combined with the cooling effect of zinc. This allows the use of higher current, increasing the energy level of the EDM spark and, as a result, the metal removal rate.

MCU

The MCU for the wire-cut EDM can be separated into three individual operator panels. Although some of the newer wire-cut EDMs eliminate some of these controls and incorporate them as part of the machine's automatic cutting cycle, a knowledge of what is being controlled during the cutting cycle should give the operator a better overall understanding of the wire-cut machine and cutting process.

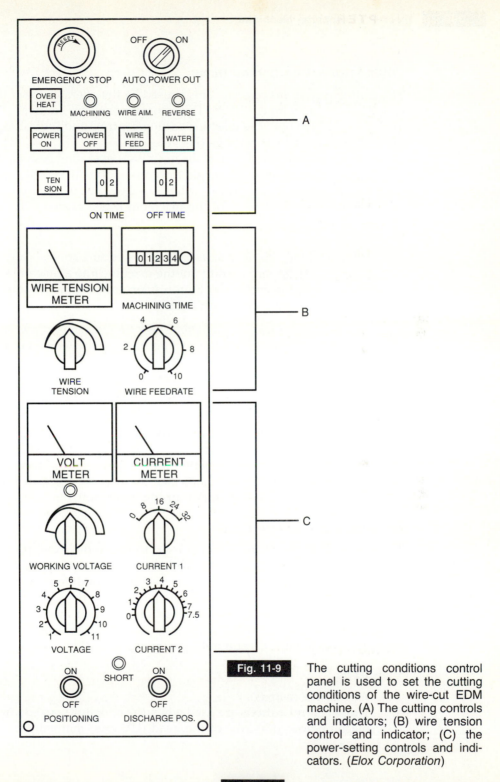

Fig. 11-9 The cutting conditions control panel is used to set the cutting conditions of the wire-cut EDM machine. (A) The cutting controls and indicators; (B) wire tension control and indicator; (C) the power-setting controls and indicators. (*Elox Corporation*)

Cutting Conditions Control Panel

The cutting conditions control panel is used by the operator to set the cutting conditions of the wire-cut EDM. This control panel can be divided into general areas of operation: the machine cutting controls and indicators, the wire tension, and the power settings (Fig. 11-9).

These controls are used to start the machine's operating systems before the cycle start button is pressed to begin the cutting operation. When the cutting cycle is in progress, these controls will be used by the operator to monitor and adjust the cutting conditions of the machine, if necessary.

Main Parts

1. *Machine cutting controls and indicators:* The cutting controls and indicators, (Fig. 11-9A) are used to put the machine into a machining-ready state. When the cycle start button is pressed on the machine setup panel, the cutting cycle will start.

2. *Wire tension controls and indicator:* The wire tension controls and indicator (Fig. 11-9B) are used to set and monitor the tension of the wire (electrode), which is measured by a voltmeter. The voltage reading represents the intensity of the brake or pull that is being used to apply the tension to the wire. The wire feed rate controls the speed of the wire as it passes through the cutting area of the workpiece.

3. *Power setting and indicators:* The two sets of controls in Fig. 11-9C are used to set and monitor the voltage and current. The voltmeter indicates the voltage supplied between the workpiece and the wire (electrode). The current meter indicates the current between the workpiece and the wire (electrode) during machining.

Machine Setup Control Panel

The machine setup control panel is used to locate the exact position of the wire (electrode) in relation to the workpiece, and it stores these coordinate locations. The operator also sets the required conditions and operations needed to read the program accurately and cut the workpiece. This control panel can be divided into general areas of operation and control: the axis motion controls, mode position control, feed controls, functional controls, reference and indicators, and auto power recovery (Fig. 11-10).

Main Parts

1. *Axis motion controls*
   These control buttons are used when the operator wants to move the machine table (XY axis) or the head (UV axis) manually.

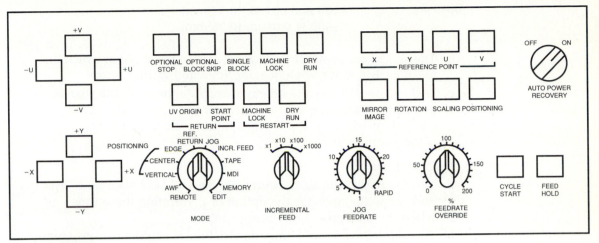

Fig. 11-10 The machine setup control panel is used to locate the position of the wire (electrode). (*Fujitsu Fanuc, Ltd.*)

2. *Mode position control*

   This switch sets the operation mode selected by the operator.

   a. The *edit, memory, MDI,* and *tape mode* are used to perform these respective functions when the operator is working with the part program.

   b. The *incremental feed and jog mode* are used when positioning the machine manually.

   c. *Reference return mode* is used when the operator is returning the machine to its reference (or home) position.

   d. The *edge, center,* and *vertical positioning* are used when locating the position of the wire (electrode).

   e. *Automatic wire feed (AWF)* is used when the wire (electrode) is to be threaded automatically.

   f. *Remote* is used if the machine is equipped with a remote pendant, which gives the operator the ability to activate the machine controls without having to use the machine setup control panel.

3. *Feed controls*

   a. The *incremental feed control* switch sets the incremental feed of the machine.

$$\text{X1 is 0.00001 in. (0.0002 mm)}$$
$$\text{X10 is 0.0001 in. (0.002 mm)}$$

X100 is 0.001 in. (0.02 mm)
X1000 is 0.01 in. (0.25 mm)

  b. The *jog feed rate* allows the operator to select the speed at which the machine will move while jogging (moving) during setup.

  c. *Feed rate override* lets the operator override the feed rate that was put in the part program. The programmed feed rate can be increased or decreased during the cutting cycle to help stabilize the cutting condition.

4. *Functional controls*

  a. The *functional controls, optional stop, optional block skip,* and *single block* allow the operator the option of performing these functions when the part program is being run.

  b. The *machine lock switch* allows the machine to run the part program and display it on the CRT without any machine movement.

  c. The *dry run switch* allows the operator to run the part program automatically without actually cutting the part. In this mode, the operator can draw out on paper the machine's cutting path.

  d. The *reference indicators, XYUV reference points, mirror image, rotation, scaling,* and *positioning* are indicators that light up when the machine has performed any of these functions or is set to perform these functions.

5. *Auto power recovery*
The auto power recovery switch is used during unattended machining. If the power supply to the machine is interrupted, the machine will automatically restart as soon as the power comes on and will continue the cutting operation until the workpiece has been completed.

MDI AND DISPLAY PANEL

The MDI and display panel (Fig. 11-11) are used by the operator to put information or a program into the memory of the MCU manually. The CRT can display machine parameters (XYUV locations), maintenance diagnostics, alarms, the entire program for editing purposes, or the program while the part is being machined. The control panel can be divided into general areas of operation: the CRT screen, cursor and page buttons, address and data keys, and function buttons.

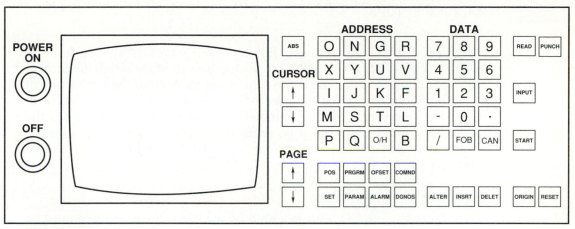

Fig. 11-11   The MDI and display panel. (*Fujitsu Fanuc, Ltd.*)

Main Parts

1. *CRT Screen*
The CRT screen used in MCUs can be either monochrome or color. It displays to the operator information that is stored in memory or taken from paper tape or other storage media.

2. *Cursor and Page Buttons*
The cursor buttons allow the operator to control the location of the cursor on the CRT screen and to input or edit information that is displayed.

- The page buttons allow the operator to move from one page to another of a part program, just as a person would turn the pages of a book when reading.

3. *Address and Data Keys*
The address and data (numeric) keys are used to input the individual characters and/or numbers needed to create a part program, edit existing programs, or change the machine's parameters to achieve different machining conditions.

4. *Function Buttons*
The function buttons display the program, offsets, diagnostics, parameters, etc. An operator uses the function buttons just as a reader turns from chapter to chapter in a book.

MACHINE SETUPS

Most work machined on a wire-cut EDM is bolted to the U-shaped table of the machine (Fig. 11-12). The table has two stop bars used for rough alignment. They are adjustable or can be removed if required. Around the top surface of the table are equally spaced threaded holes that are used to clamp the work-piece. The workpiece must be accurately located square and parallel to the table. The wire electrode must be set to the required relationship with the workpiece to enable the operator to produce a part that is dimensionally accurate.

Alignment of the Workpiece

Once the workpiece has been mounted on the table, the workpiece must be aligned with the table travel. The two most commonly used reference points are an edge and a hole. The wire-cut EDM has the ability to accurately locate an edge or a hole, but the operator must first locate the workpiece. The work-piece can be located in a number of ways, but it is the skill of the operator in performing this setup that will determine the final outcome of the part.

Threading the Wire

Once the correct wire type and size for the job have been selected, the opera-tor must mount the spool of wire electrode and thread the wire in the proper sequence. The exact path that the wire will follow will be determined by the

Fig. 11-12 The wire-cut EDM work table and workpiece. (*Fujitsu Fanuc, Ltd.*)

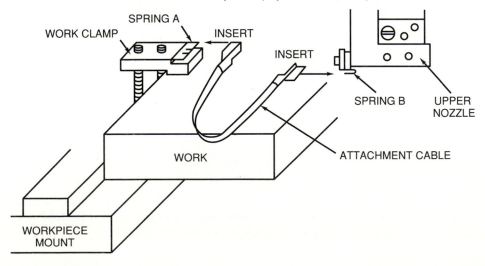

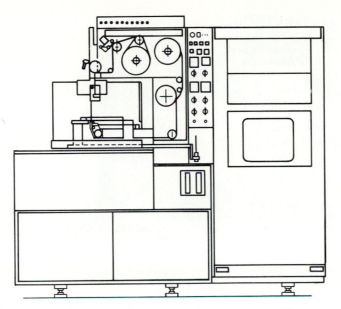

Fig. 11-13 A typical wire thread diagram from the supply spool through the workpiece to the takeup reel. (*Fujitsu Fanuc, Ltd.*)

make and model of the wire-cut EDM machine (Fig. 11-13). The wire must run freely around the rollers and through the guides (Fig. 11-14). If the wire is not correctly threaded and touches the body of the machine, a short circuit will occur when the power to the wire is turned on. This "short condition" will prevent any machining from taking place. The machine will shut off and indicate that the short must be corrected before the machine can begin a cutting cycle.

Vertical Alignment of Wire

Before a workpiece is machined, the wire (electrode) must be aligned vertically. If the wire is not in a vertical position, the operator cannot machine an accurate part even if the program for the workpiece is correct. A granite setup block with metal contacts is used to vertically align the wire (Fig. 11-15).

The setup block is clamped to the worktable with the metal contacts facing the cutting area and then plugged into the MCU. As the wire (electrode) touches the contact points, the MCU receives an input signal and adjusts the U and V axes head until the wire has been vertically aligned. During this operation, all operating switches on the cutting conditions control panel should be set as low as possible, the wire should be moving, and the wire tension on.

Fig. 11-14   The upper and lower wire guides keep the wire (electrode) in a vertical position. (*LeBlond-Makino Machine Tool Company*)

Wire Tension

After the wire has been threaded on the machine, the correct wire tension must be set with a wire tension gage (Fig. 11-16). When the tension switch is in the ON position, the amount of tension set by the tension dial is read on the current meter as electrical current that has been applied to the tension brake. The amount of tension that is used during the machining operation will vary with the different types and diameters of wire (Table 11-1). The tension applied to the wire is constantly being monitored by a wire break detecting lever. If the wire breaks during the cutting cycle, this lever will send a signal to the MCU and the cutting cycle will automatically stop. The machine will enter into a feed-hold condition until the wire is rethreaded either manually or automatically and the cycle start button is pressed to continue the cutting cycle.

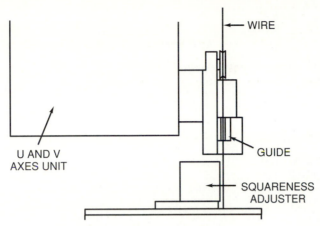

WIRE

GUIDE

U AND V
AXES UNIT

SQUARENESS
ADJUSTER

Fig. 11-15 Vertical alignment of the wire (electrode) using a setup block. (*Fujitsu Fanuc, Ltd.*)

AUTOMATIC EDGING AND CENTERING

When the workpiece is set up for machining, it is sometimes necessary to locate the wire either at an edge of the workpiece or in the center of a hole. This can be done automatically if the control unit is equipped with an edging function or a centering function.

Fig. 11-16 Wire tension gage (tensiometer) for gaging the correct wire tension. (*LeBlond-Makino Machine Tool Company*)

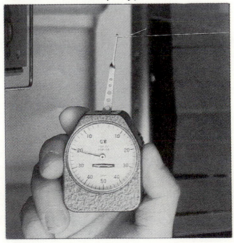

Table 11-1

| Diameter of wire electrode | | Wire tension | | | |
| | | Brass, molybdenum, zinc coated | | Copper | |
in.	mm	oz	g	oz	g
0.002	0.05	7	200	3.5	100
0.003	0.07	10.5	300	5.3	150
0.004	0.10	14	400	7	200
0.005	0.12	17.6	500	8.8	250
0.006	0.15	21.2	600	10.5	300
0.007	0.17	24.7	700	12.3	350
0.008	0.20	28.2	800	14	400
0.009	0.22	31.7	900	15.8	450
0.010	0.25	35.2	1000	17.6	500
0.011	0.27	38.8	1100	19.4	550
0.012	0.30	42.3	1200	21.2	600

Automatic Edging

The *automatic edging function* is used when it is necessary to locate the wire at the edge of a workpiece (XY zero). The edge of the workpiece is located when the wire moves over and touches the edge of the workpiece. A signal is generated by the short circuit between the wire and the workpiece, and this procedure is repeated several times. The detected edge locations are then averaged by the MCU to give an accurate edge location. The machine then automatically positions the wire at this point (Fig. 11-17).

Locating the Edge of a Workpiece

1. Set up the workpiece on the machine table.

2. Set the MCU to the proper mode (edge find).

3. Apply tension to the wire by turning on the tension switch (the wire should be running).

4. Set the feed rate selection to X1 (0.00001 in., or 0.0002 mm) or X10 (0.0001 in., or 0.002 mm).

5. Press the axis button to move the wire toward the edge surface.

6. The machine start light will indicate that the edge finding has started.

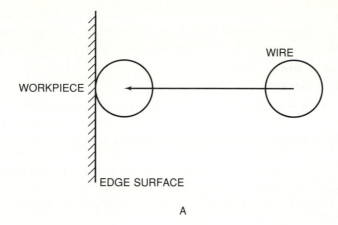

A

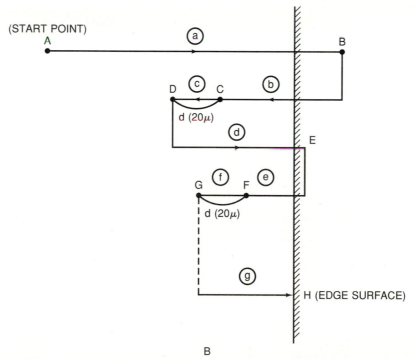

B

Fig. 11-17 The automatic edge-finding sequence is used to locate the edge of the workpiece. (*Fujitsu Fanuc, Ltd.*)

7. When the edge finding is completed, the start light will go off and the position light will come on, indicating that the edge finding is completed.

Automatic Centering

The automatic centering function is used to locate the wire (electrode) in the center of a hole. Like edge finding, the side of the hole is detected by the wire touching the workpiece. A short circuit is created between the wire and the workpiece along the X axis, sending a signal to the MCU indicating the side of the hole. The wire then moves to the opposite side of the hole until a side is detected. After the second side has been detected, the wire moves toward the center of the hole and performs the same function along the Y axis (Fig. 11-18).

Locating the Center of a Hole

1. Set up the workpiece on the machine table.

2. Thread the wire through the machined hole so that it is approxi-

Fig. 11-18 The automatic cycle for rough-centering the wire in a hole. (*Fujitsu Fanuc, Ltd.*)

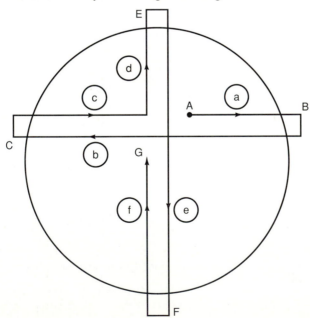

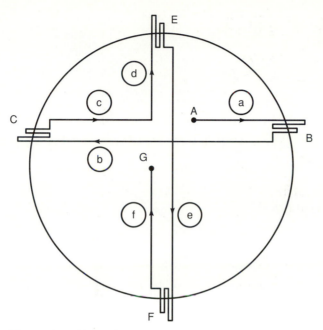

The automatic cycle accurately centers the wire after the rough center has been performed. (*Fujitsu Fanuc, Ltd.*)

mately in the center. A short circuit condition cannot exist between the wire and the workpiece at this time.

3. Set the MCU to the proper mode (centering).

4. Apply tension to the wire by turning on the tension switch (the wire should be running).

5. Set the feed rate selector switch to X1 (0.00001 in., or 0.0002 mm) or X10 (0.0001 in., or 0.002 mm).

6. Press the axis button to move the wire in either the X or the Y axis.

7. The machine start light will come on, indicating that centering has started.

8. When the centering is completed, the start light will go off and the position light will come on, indicating that the centering is completed.

The machine performs a rough centering first to locate the wire at the approximate center before accurately centering the wire (Fig. 11-19).

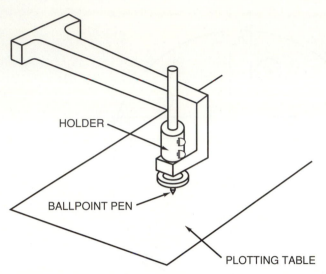

HOLDER

BALLPOINT PEN

PLOTTING TABLE

Fig. 11-20 Drawing equipment to plot the outline of the programmed part during the dry run or machining cycle. (*Fujitsu Fanuc, Ltd.*)

Dry Run

When the wire machine has been set up, and before the actual cutting operation, the operator can check the tape program for accuracy by using the *dry run mode*. This mode allows the operator to run the part program without actually cutting a part. If the wire machine is equipped with drawing equipment (Fig. 11-20), the outline of the programmed part can be plotted on paper as the machine goes through the complete program. During the dry run cycle, the machine is usually set to move at rapid travel; however, the rate of travel can be controlled by the jog feed rate switch.

Test Square

The test square is a program that is run to cut a 0.100-in. (2.54-mm) square before cutting the actual part (Fig. 11-21). The operator can use this square to check for dimensional accuracy of the cut and, if necessary, allow for more or less wire diameter compensation to produce the correct size. The surface finish that will be obtained during the cutting operation can be checked, and any adjustment can be made to change the cutting conditions. Once the operator is satisfied with the accuracy and finish obtained, the part program can be run and the desired workpiece produced.

Skim Cutting

Skim cutting on the wire-cut EDM is similar to taking a finish cut with a conventional machine tool. Once most of the material has been removed, the operator can use MDI to offset a wire diameter into the MCU by 0.0005 in. (0.012 mm). The voltage, current, and cutting speed can be increased for this operation. When the adjustments to the cutting conditions have been made, the part program is run again, skimming the programmed part boundary, producing a smooth finish and an accurately sized part.

Note: If the reverse light or the low-voltage light comes on during the skim cutting operation, the cutting speed is too fast. If the cutting operation is

Fig. 11-21   An incremental and absolute program for a test square to check the cutting conditions and the part size *before* the workpiece is machined.

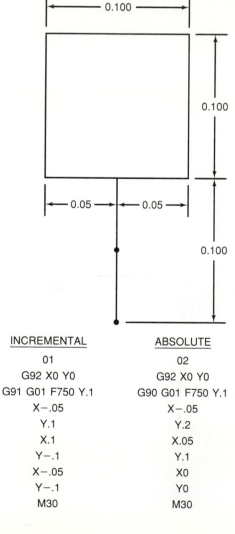

INCREMENTAL	ABSOLUTE
01	02
G92 X0 Y0	G92 X0 Y0
G91 G01 F750 Y.1	G90 G01 F750 Y.1
X−.05	X−.05
Y.1	Y.2
X.1	X.05
Y−.1	Y.1
X−.05	X0
Y−.1	Y0
M30	M30

skipping but not reversing, increase the cutting speed. The skim cutting operation can also be used to put small tapers (¼° to ½°) on the workpiece after the initial shape has been cut.

Taper Cutting

Taper cutting is performed by the simultaneous control of the upper wire guide (the U and V axes) as well as the lower wire guide (the X and Y axes). This sets the wire (electrode) to the desired angle from the vertical position (Fig. 11-22).

When the wire inclination code left (G51) or wire inclination code right (G52) is used, the upper guide is moved to the right or left of the programmed path. The amount of taper that can be programmed depends on the type of machine tool being used. If an angle is required that is greater than the machine's capability, the workpiece can be mounted on an angle to cut larger angles.

Fig. 11-22 The upper and lower wire guides can be offset for cutting tapered sections. (*LeBlond-Makino Machine Tool Company*)

Taper Corner Cutting

This function is used to produce arcs at corners when an oblique cylindrical surface is required (Fig. 11-23), rather than the conical surface that is formed when arcs are programmed at corners (Fig. 11-24).

RECOMMENDED CUTTING CONDITIONS

The recommended cutting conditions (Table 11-2) are used by the operator when setting up the machine and the cutting conditions control panel. It is recommended that a similar chart be set up by the operator each time a cutting operation is performed, using different combinations of electrode material and workpiece material. A library can be quickly built up with the results obtained from the different types of electrode, workpiece materials, and machine settings. This will prove to be an invaluable reference resource as well as a timesaver during the setup and cutting operations.

The acceptance of EDM is increasing as more applications are found for this process. As new and important technological advances in equipment and application techniques become available, more industries will adopt the EDM process. When metal removal rates become more comparable with conventional metal removal processes, the advantages of EDM will make it very difficult for industry to justify traditional machining techniques.

PROGRAMMING A WIRE-CUT EDM MACHINE

The wire cut exercise shown in Fig. 11-25 on page 323 applies all the theory of this entire section to cut the anchor design.

Notes:

1. The workpiece material is 1-in.-thick tool steel.

2. The wire is brass and is 0.008 in. in diameter.

3. See Table 11-2 for recommended cutting conditions and set the machine controls as necessary.

4. Fasten the workpiece at any convenient spot on the table.

5. Sections to be cut away should be secured (glued) to prevent them from shorting out the machine as the section separates from the main part.

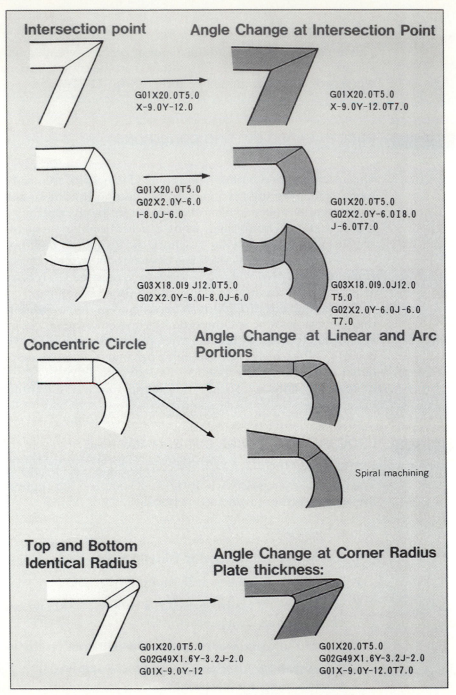

Fig. 11-23 Oblique cylindrical corner. (*LeBlond-Makino Machine Tool Company*)

Fig. 11-24 Conical corner formed when arcs are programmed. (*LeBlond-Makino Machine Tool Company*)

Table 11-2 RECOMMENDED CUTTING CONDITIONS

Workpiece material		Die steel hardened	Die steel hardened	Tungsten carbide	Aluminum	Copper	Graphite
Wire material	(electrode)	Brass annealed	Brass annealed	Brass annealed	Brass annealed	Brass annealed	Brass annealed
Wire diameter	(in.)	0.008	0.008	0.008	0.008	0.008	0.008
	(mm)	0.20	0.20	0.20	0.20	0.20	0.20
Workpiece thickness	(in.)	1	4	¾	⅜	1¼	¾
	(mm)	25	100	19	9.50	32	19
1. Voltage	(TAP)	1	2	3	4	3	3
2. Peak current	(TAP)	39.5	39.9	17	6.5	16.5	13
3. Capacitance	(μF)	2.0	1.5	1.5	1.5	2.0	1.5
4. ON time/OFF time	(μsec)	6:5	6:5	6:5	6:5	6:5	6:5
5. Tension	(oz)	15.8	14	17.6	17.2	17.6	17.2
	(g)	450	400	500	490	500	490
6. Wire feed indication		6	8	6	5.5	8	6.5
7. Command feed rate	(in.)	0.100	0.020	0.035	0.310	0.041	0.041
	(mm)	2.50	0.51	0.90	8.00	1.05	1.05
8. Resistivity of water	($\times 10$ cm)	2–3	2–3	2–3	5	2–3	2–3
9. Actual cutting speed	(in.)	0.100	0.020	0.035	0.240	0.041	0.041
	(mm)	2.50	0.051	0.90	6.00	1.05	1.05
10. Average working voltage	(V)	58	62	75–80	75–80	80–85	95
11. Average working current	(A)	7.7	7.2	4.8–5.0	2.7	4.2–4.4	2.6
12. Surface roughness	(μm)	20–22	16–18	13	20–25	14	5
13. Cutting gap	(in.)	0.010	0.013	0.100	0.011	0.013	0.012
	(mm)	0.25	0.33	0.25	0.28	0.33	0.30
14. Distance between upper nozzle and workpiece	(in.)	0.039	0.039	0.590	0.590	0.590	0.590
	(mm)	1.00	1.00	15.00	15.00	15.00	15.00
15. Flushing rate—upper and lower nozzles		5.5	5.5		3.4		

The wire EDM program uses the following codes and commands:

1. *Word address:* The address which is an alphabetic symbol (A–Z) and a numerical value.

2. *Input format:* Words which make up a block of information must be commanded in a predetermined format. The format specifies the variation in the number of words in a block and the number of characters permitted. The format used for the wire EDM program is shown in Table 11-3.

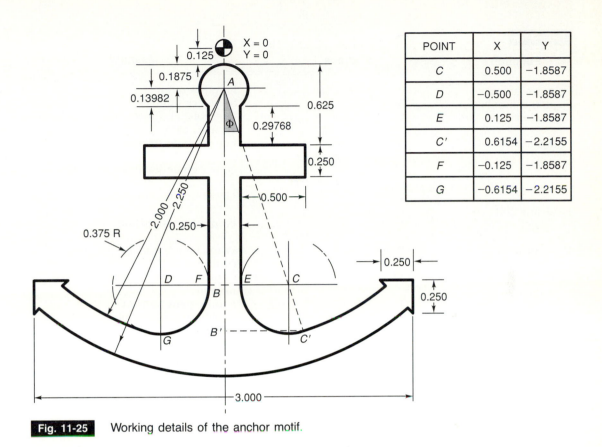

POINT	X	Y
C	0.500	−1.8587
D	−0.500	−1.8587
E	0.125	−1.8587
C′	0.6154	−2.2155
F	−0.125	−1.8587
G	−0.6154	−2.2155

Fig. 11-25 Working details of the anchor motif.

3. *Inch and metric input:*

N G X Y I J (U V) F T M

4. *Miscellaneous functions:* The EDM wire-cut miscellaneous functions have slight variations from those used on machining centers:

M00	Program stop
M01	Optional program stop
M02	Program end
M30	Program end and rewind

Table 11-3

Function	Address		Meaning
Program number	0	1–9999	Program number
Sequence number	N	1–9999	Sequence number
Preparatory function	G	00–99	Interpolation (linear, circular, etc.)
Coordinate words	XY)	) in. 999.99999	Lower guide traverse command
	UV)		Upper guide traverse command
		) mm 99999.999	
	I J)		Lower guide circle center coordinates
Feed function		in. 0–1.0000	
	F		Feed rate designation
		mm 0–25.00	
Wire inclination	T	± 5.0°	Wire angle command for taper cutting
Miscellaneous functions	M	0–99	Control commands for machine
Subprogram	P	0–9999	Subprogram number
	L	0–99	Repetition number of subprogram

M40	Discharge off
M80	Discharge on
M98	Subprogram call
M99	Subprogram end

5. *Preparatory functions:* The preparatory functions for an EDM wire-cut machine differ slightly from similar functions used for machining centers:

G Code	*Preparatory Function*
G00	Rapid traverse
G01	Linear interpolation
G02	Circular interpolation (CW)
G03	Circular interpolation (CCW)
G04	Dwell
G17	Plane selection (XY plane)
G20	Inch data input

G Code	Preparatory Function
G21	Metric data input
G28	Automatic zero return
G30	Start point return
G40	Wire diameter compensation cancel
G41	Wire diameter compensation left
G42	Wire diameter compensation right
G50	Wire inclination cancel
G51	Wire inclination left
G52	Wire inclination right
G90	Absolute programming
G91	Incremental programming
G92	Coordinate system setting

The Anchor Program: **(Fig. 11-25)**

```
%
```
Rewind/stop code.

```
N010 O55
```
Program number.

```
N020 [ANCHOR]
```
Program name.

```
N030 G92 X0.0 Y0.0
```
G92–position preset.

Moves from table home position to program part XY zero.

```
N040 G41 F75 G90 G01 Y − 0.125
```
G41—wire offset to right.

F75—feed rate (in./min).

G90—Absolute program.

```
N050 G02 X0.125 Y − 0.4523 I0 J − 0.1875
```
G02—circular interpolation clockwise (CW).

I and J—values for arc location.

N060 G01 X1.25 Y − 0.750
G01—linear move in the Y direction of 0.750 in.

N070 X0.625   )
)
N080 Y − 1.0   )
)
N090 X0.125   )
)
N100 Y − 1.8587   )
Linear moves.

N110 G03 X0.6154 Y − 2.2155 I0.375 J0.0
G03—circular interpolation counterclockwise (CCW).

N120 G03 X1.2866 Y − 1.8437 I − 0.6154 J1.9030
I and J—values for arc location.

N130 G01 X1.250 Y − 1.8177   )
)
N140 X1.500   )
)
N150 Y − 2.0677   )
Linear moves.

N160 X1.4634 Y − 2.0311

N170 G02 X − 1.4634 Y − 2.0216 I − 1.4634 J1.7091
G02—circular interpolation clockwise (CW).

N180 G01 X − 1.500 Y − 2.0677   )
)
N190 Y − 1.8177   )
)
N200 X − 1.250   )
)
N210 X − 1.2866 Y − 1.8543   )
Linear moves.

```
N220 G03 X − 0.6154 Y − 2.2155 I1.2866 J1.5462    )
                                                  )
N230 X − 0.125 Y − 1.8587 I0.1153 J0.3568         )
```
    G03—circular interpolation counterclockwise (CCW).

```
N240 G01 X − 0.125 Y − 1.000   )
                               )
N250 X − 0.625                 )
                               )
N260 Y − 0.750                 )
                               )
N270 X − 0.125                 )
                               )
N280 Y − 0.4523                )
```
    Linear moves.

```
N290 G02 X0.0 Y − 0.125 I0.125 J0.1398
```
    G02—circular interpolation clockwise (CW).

```
N300 G01 X0.0 Y0.0 G40
```
    Return to program start position.
    G40—cancel wire offset.

```
N310 M30
```
    End of program.

```
N320 %
```
    Rewind/stop code.

REVIEW QUESTIONS

EDM

1. Define *EDM*.

2. What types of materials are especially suited for machining by the EDM process?

Principles of EDM

3. Name two types of EDM machines used in industry today.

4. How many axes can be controlled at the same time?

Wire-Cut EDM Parts

5. Briefly state the purpose of the following wire-cut EDM parts: bed, saddle, table, column, wire feed system, dielectric system, MCU.

The Servomechanism

6. What does the EDM power supply control?

7. What is the approximate size of the spark gap between the wire and the workpiece?

8. Why is it important that the electrode and the workpiece not touch during the cutting operation?

9. How is machining affected by:
   (a) Too large a spark gap?
   (b) Too small a spark gap?

The Dielectric Fluid

10. Explain the effect of poor flushing on the cutting operation.

11. What type of dielectric fluid is generally used in a wire-cut EDM?

12. How is the amount of deionization measured?

13. How is the water flow to the workpiece controlled?

14. What problem exists if red sparks occur during the cutting operation?

The Electrode

15. What is the range of electrode wire diameters?

16. Why does electrode wear not affect the accuracy of the cut?

17. What characteristics should a good electrode material possess?

18. Name the most common materials that are used in the manufacture of wire electrodes.

19. What are two characteristics of tungsten and molybdenum wire?

20. Define *stratified wire* and state how it increases the metal removal rate.

MCU

21. What is the purpose of the cutting conditions control panel?

22. What part of the machine setup control panel is used to control the table movement and the feed rate?

23. What is the purpose of the manual data input and display panel?

Machine Setups

24. How is a workpiece generally held for cutting?

25. What are two commonly used reference points for setup?

26. What could cause a "short condition" when threading the wire?

27. How does a "short condition" affect the cutting operation?

28. Why is it important that the wire be in a vertical position for cutting?

29. How is the wire (electrode) set to a vertical position?

30. At what settings should the operating switches be set during vertical alignment of the wire (electrode)?

31. How is the wire tension indicated?

32. How is the wire tension set?

33. How is the wire tension monitored on the machine?

Automatic Edging and Centering

34. How may the edge of a workpiece be accurately located?

35. Briefly describe the operation of locating the wire in the center of a hole.

36. What is the purpose of the dry run mode?

37. What is the purpose of cutting a test square?

Skim Cutting

38. Define *skim cutting*.

39. How does the voltage, current, and cutting speed differ when skim cutting?

40. What would indicate that the cutting speed is too fast or too slow when skim cutting?

Taper Cutting

41. How is taper cutting performed?

42. What do the following codes represent?
(a) G51
(b) G52

43. How could a taper that is greater than the machine's capability be cut?

CHAPTER

TWELVE

Numerical Control and the Future

Numerical control (NC) is a technology that combines electronic hardware with software programs to perform various operations in the machine tool industry. At first, NC instructions or programs were manually prepared; however, it soon became apparent that manual programming was too slow and cumbersome for complex part geometries, and as a result, various programming languages were developed to make this task easier. Today, NC programming is assisted by computer-aided design (CAD), where the geometry of a complex CAD component can be translated into NC program instructions. Programs can also be made by tracing solid models or patterns of an object to produce a NC program. It is quite likely that NC systems will play a very important role in joining engineering and manufacturing information for the factory of the future. Therefore, the stand-alone NC systems of the past are fast becoming an important component in intelligent manufacturing cells and the factories of the future.

1. Understand the role of NC and CAD in the machine tool industry

2. Understand how CAD and computer-assisted manufacturing (CAM) make interactive programming possible

3. See how NC will play an ever-increasing role in the factory of the future

PROGRAMMABLE AUTOMATION

During the past ten years, CAM and the automated or unattended factory has presented engineers with a real challenge. The prospect of precise control and accuracy of the manufactured part and the increased productivity which is possible through full implementation of NC are creating great interest in manufacturing industries.

The development of NC and the great advances in computer technology have resulted in what is commonly described as the *second industrial revolution*. This is rapidly progressing toward the *factory of the future*, where the entire operation of a complete manufacturing plant can be programmed and controlled by a communication network that joins engineering data design with manufacturing data and the automated machines on the factory floor. Much of the technology required for a computer-based manufacturing plant exists today.

1. Computers and communication networks are capable of controlling the operations of an entire plant.

2. Machine tools and assembly equipment which respond to computer instruction are a reality.

3. CAD and computer-aided engineering (CAE) systems can be linked to the factory floor.

4. Manufacturing requirements planning (MRP) systems are available to achieve manufacturing control.

The factory of the future will become a reality as soon as the various components can be joined together so that coordination, information transfer, and broad-based planning can occur. A new manufacturing architecture must be planned where each component of a CAM system is arranged in the proper order or class, each of which is subject to or dependent upon the one above or below it. This architecture or sequence must include all the levels shown in Fig. 12-1.

Process equipment: Conventional or nonprogrammable machine tools may include some high-speed transfer machines.

Fig. 12-1 The logical arrangement or order of equipment from conventional (standard) machines to the factory of the future.

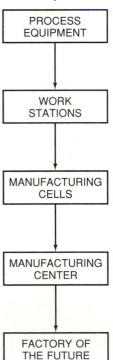

PROCESS
EQUIPMENT

WORK
STATIONS

MANUFACTURING
CELLS

MANUFACTURING
CENTER

FACTORY OF
THE FUTURE

Workstations: These include computers, NC machines, and robots.

Manufacturing cell: A group of machines or workstations would be combined with material handling equipment to produce a subassembly or major component.

Manufacturing center: This consists of a group of manufacturing cells joined together to produce a major subassembly.

Factory of the future: All the levels would be integrated (joined) with central computer controls for process and capacity planning, manufacturing operations, quality control, product tracking, and engineering data.

Since no factory of the future can become a reality without components such as NC, direct numerical control (DNC), CAD, CAM, etc., it may be advisable to examine each more fully.

DNC

DNC was introduced in the mid-1970s as a method of overcoming some of the shortcomings of earlier NC systems such as programming errors, punched tape input, limited control unit functions, and a lack of operational data. DNC, where a computer is linked directly to the machine control unit (MCU), can load programs for many machine tools from a central computer and eliminate the need for punched tape or tape readers (Fig. 12-2).

Fig. 12-2   The main parts of a DNC system.

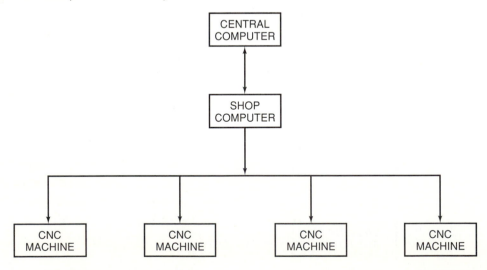

DNC provided a means whereby a part program could be distributed to one or more NC/CNC machine control units, and the program could then be stored, edited, executed, and returned to the main computer. Today DNC implies that information is being transferred from one computer system or machine control to another, and might better be referred to as *distributed numerical control.*

COMPUTER NUMERICAL CONTROL (CNC)

CNC, a direct offshoot of DNC, resulted from the invention of the minicomputer. In CNC, a single dedicated computer or minicomputer, located on or near a machine, is used to control the operations of a *single NC machine tool.* The minicomputer, generally built into or beside an MCU, can provide the same advantages as a DNC system without being dependent upon the central or host computer. The CNC unit contains many features which make it very popular for small or specialized machining operations (Fig. 12-3). Some of the more important features are:

1. *Memory*
   The details of machining a part can be stored in the memory either by punched tape or manually by manual data input (MDI). Once the program is in memory, the data can be recalled whenever it is necessary to produce a part.

2. *Diagnostics*
   This provides troubleshooting features (diagnostic tests) to identify and correct the problem when a CNC unit does not operate ("goes down"). There are generally two methods used for diagnosing problems:
   a. A special diagnostic tape supplied with the CNC unit, which checks many different elements and displays the information on the CRT screen by means of signal lights.
   b. By a telephone line connection to the CNC manufacturer, who can run tests and spot the problem.

3. *Special routines*
   Common routines, such as bolt hole circles, and pocketing routines can be computed from a simple descriptive statement.

4. *Electronics Industries Association (EIA) vs. American Standard Code for Information Interchange (ASCII)*

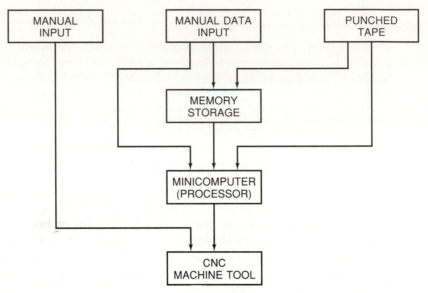

Fig. 12-3 A CNC unit operates through its minicomputer.

Most newer CNC machine control units can read either EIA or ASCII code standards, which are identified through parity check.

5. *Inch vs. metric*
Most CNC control units can handle both inch and metric measurements.

Many CNC systems contain applications software programs which provide canned cycle capabilities, permitting a machine operator to recall a stored program by a few keystrokes on the control panel. Macros for machine motion, tooling, and other machine functions can be performed with suitable applications software. The success of any CNC system is largely dependent on the capabilities provided by the software programs that are available to control it.

CAD

The advent of the computer proved to be a boon to the design engineer in that it simplified the long, tedious calculations which were often involved in designing a part. In 1963, the Massachusetts Institute of Technology (MIT) demonstrated a computer system called Sketchpad that created and displayed graphic information on a cathode-ray tube (CRT) screen. This system soon became known as CAD, and it allows the designer or engineer to produce finished engineering drawings from simple pencil sketches or from models and modify these drawings on the screen if they do not seem functional. From three-drawing, orthographic views, the designer can transform drawings into a three-dimensional view and, with the proper computer software, show how the part would function in use. This enables a designer to redesign a part on the screen, project how the part will operate in use, and make successive design changes in a matter of minutes.

CAD Components

CAD is a televisionlike system that produces a picture on the CRT screen from electronic signals received from a computer. Most CAD systems consist of a desk-top computer which is connected to the main or host computer. The addition of a keyboard, light pen, or an electronic tablet and plotter enables the operator to produce any drawing or view required (Fig. 12-4).

The operator generally starts with a pencil sketch and, with the use of the light pen or an electronic tablet, can produce a properly scaled drawing of the part on the CRT screen and also record it in the computer memory. If design changes are necessary, the designer is able to create and change parts and lines on the CRT screen with a light pen, an electromechanical cursor, or an electronic tablet (Fig. 12-5).

After the design is finished, the engineer or designer can test the anticipated performance of the part. Should any design changes be necessary, the engineer or designer can make changes quickly and easily to any part of the drawing or design without having to redraw the original. Once the design is considered correct, the plotter can be directed to produce a finished drawing of the part.

CAD offers industry many advantages which result in more accurate work and greater productivity. The following is a list of some of the more common CAD advantages:

- Greater productivity of drafting personnel
- Less drawing production time
- Better drawing revision procedures
- Greater drawing and design accuracy
- Greater detail in layouts
- Better drawing appearance
- Greater parts standardization
- Better factory assembly procedures
- Less scrap produced

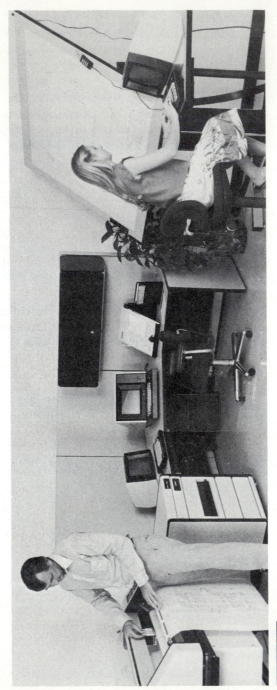

Fig. 12-4 A total CAD system. (*Bausch & Lomb*)

Fig. 12-5   An electronic tablet being used to make design changes on the part shown on the CRT screen. (*Bausch & Lomb*)

CAM

CAM started with NC in 1949 at MIT. This project, sponsored by the U.S. Air Force, was the first application of computer technology to control the operation of a milling machine. Figure 12-6 illustrates the CAD/CAM evolution from the 1950s to the 1980s.

Standard NC machines greatly reduced the machining time required to produce a part or complete a production run of parts, but the overall operation was still time-consuming. Tape had to be prepared for the part, editing the program would result in making a new tape, and tapes had to be rewound each time a part was completed. With this in mind, the machine manufacturers added a computer to the existing NC machine, introducing the beginning of CNC.

The addition of the computer greatly increased the flexibility of the machine tool. The parts program was now run from the computer's memory instead of from a tape that had to be rewound. Any revisions or editing of the program could be done at the machine, and changes could be stored.

As the machine tool manufacturers continued to improve the efficiency of their machines, the computer capabilities were greatly increased to programmable microprocessors, and many time-saving devices were introduced to increase the machine's cutting time and reduce downtime. Some of these machine options are automatic tool changers, parts loaders and unloaders, chip conveyors, tool wear monitors, in-process gaging, and robots—which brings us to today's machining centers (Fig. 12-7).

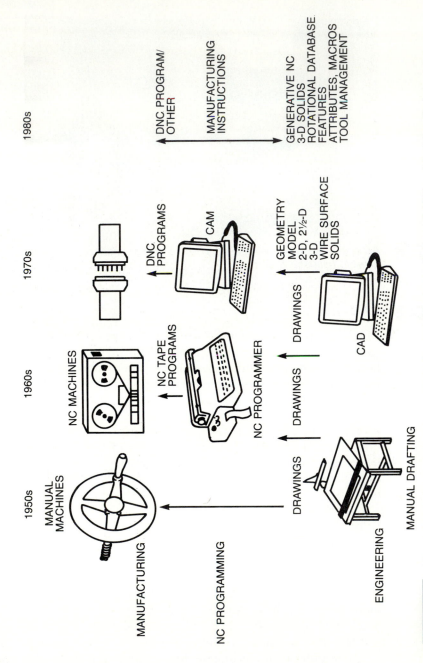

Fig. 12-6 The CAD/CAM evolution from the 1950s to the 1980s. *(Ingersoll Milling Machine Co.)*

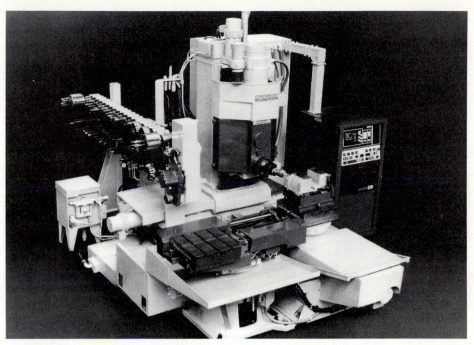

Fig. 12-7 Most machining centers are equipped with automatic tool changers and workhanding equipment in order to increase the productivity of the machine. (*Cincinnati Milacron, Inc.*)

CAM uses all the advanced technologies to automate the operations in manufacturing and handle the data that drives the process. The tools of CAM include computer technologies, CAE, and robotics. CAM uses all these technologies to join the process of design with automated production machine tools, material handling equipment, and control systems. Without computers, the most important tool in industry, the productivity of the United States would be in serious trouble. Computers help people to become more productive and to do things that would almost be impossible without them.

CAM ties together all the major functions of a factory. The manufacturing or production operations are joined together with the process planning, production scheduling, material handling, inventory control, product inspection, machinery control, and maintenance to form a total manufacturing system.

A CAM system generally contains three major divisions:

Manufacturing: The physical operation of controlling the machine tools, material handling equipment, inspection operations, etc., in order to produce the parts required

Engineering: The process which involves design and engineering activities to ensure that the parts are designed properly in order to function as required

Management: The information such as scheduling, inventory control, labor, and manufacturing costs, and all the data required to control the entire plant

CAM increases the productivity and versatility of machine tools. Before the introduction of NC and CAM, most machine tools were cutting metal only about 5 percent of the time. The automated systems available now cut metal about 70 percent of the time, and the goal is to come as close as possible to having them remove metal 100 percent of the available time.

CAD/CAM/CNC

CAD/CAM systems (Fig. 12-8) can be used to produce CNC data to machine a part. After preparing a tool list and setup plan for the required part, the CNC programmer starts by creating a database. Once this database has been created, the programmer can recall the part on the CRT screen. After the part is displayed, the programmer describes the tools required from the information in the tool library. This library contains a description and either a name or a tool identification number for each and every tool available for use. Assume that the database and tools described are for a milling/drilling operation. The next operation would then be to generate the tool path. In this case, the types of machining that could be performed would fall into three categories:

Drilling operations: Using the Z axis perpendicular with the surface of the part

Profile milling operations: Milling the profile of the part

Pocket milling operations: Using the Z axis to plunge and remove the material from the interior of the part

The CRT screen displays a graphic representation of the part and the path the cutting tools will follow in order to complete the machining of the part. This information must now be converted to a cutter location (CL) file. With a CL file, the information that was created to generate the graphic display is converted into coordinate locations to move the cutting tool around the part. This information is now in a readable format.

Since a large variety of machines and machine controllers are used by in-

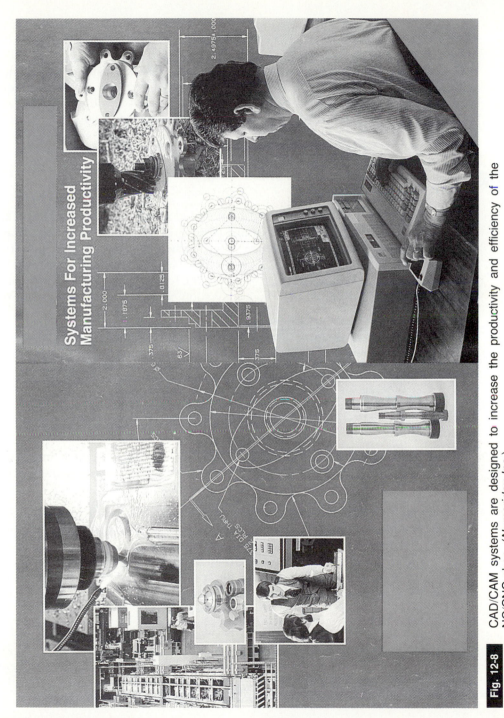

Fig. 12-8 CAD/CAM systems are designed to increase the productivity and efficiency of the NC/CNC shop. (*Numeridex Inc.*)

dustry, part processors are required. A part processor is based on the way a specific machine and machine tool controller accepts and understands NC data. There will be a postprocessor for each machine. The postprocessor takes the CL file and converts it into a tape image file. The tape image file can now be used by the CNC operator to machine the required part. To better understand what information is required and how it flows through a CAD/CAM system to generate a CNC tape image file, refer to Fig. 12-9.

NC AND CAM IN THE FUTURE

The future of the automated and unattended factory depends more on the manufacturing information programs (software) which will be available and how they will be used than on computer hardware. The factory of the future will depend on NC, which is the key. NC systems will probably improve, but no revolutionary changes are projected in the foreseeable future. The greatest change will come in MCUs which may soon contain the following features:

1. MCUs may become automatic programming devices, which keep their own program libraries.

2. MCUs will be able to communicate with other CAM system elements along a local area network (LAN) system.

3. MCUs will coordinate other slave machines, such as robots, in a manufacturing cell.

4. MCUs will have diagnostic capabilities enabling them to monitor themselves and the machines they control, to see when maintenance is required, and to report this information.

5. MCUs will have adaptive control (AC) which can provide feedback information from the cutting tool, workpiece, or machine tool so that the best machining conditions can be maintained. They may sense cutting tool temperature, vibration, a dull cutting tool, motor torque, etc., and automatically change the speeds and feeds to keep tool wear to a minimum while at the same time giving the best metal removal rates.

CNC systems of the future will be less dependent upon the part programmer because the software programs will contain an expert system for generating NC programs. The CAM systems of the future will bring together NC data

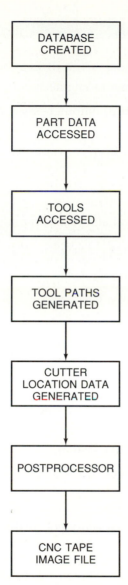

Fig. 12-9 Flow of information required by a CAD/CAM system to generate a CNC tape image file.

and other manufacturing information. The NC data combined with group technology (GT), manufacturing process planning (MPP), MRP, and manufacturing control systems will contain enough information and systems technology for the automated or unattended factory of the future.

REVIEW QUESTIONS

Programmable Automation

1. Why has CAM presented a real challenge to engineers?

2. What technology exists today to make the computer-based manufacturing plant a reality?

3. List the order or sequence of equipment from conventional machines to the factory of the future.

DNC

4. Explain how a DNC system operates and state its advantages.

CNC

5. Describe how CNC operates.

6. Briefly describe the following features of a CNC system:
   (a) Memory
   (b) Diagnostic
   (c) Special routines

CAD

7. Why has CAD become so popular and valuable to the designer or engineer?

8. Briefly describe each component of a CAD system.

9. List five of the most important advantages of CAD.

CAM

10. Trace the development of CAD/CAM from the 1950s to the 1980s.

11. List the three main tools of CAM.

12. How does CAM tie together all the major functions of a factory?

13. Name and describe the three major divisions of CAM.

CAD/CAM

14. What is a tool list or library?

15. Describe a CL file and state how it is used.

16. What purpose does the postprocessor serve?

17. List the seven steps how information flows through a CAD/CAM system.

NC and CAM in the Future

18. What is the key to the factory of the future?

19. What changes are foreseen for the MCUs of the future?

20. Why will CNC systems of the future be less dependent upon the part programmer?

21. List some of the features that will be combined with NC data for the factory of the future.

Glossary

A axis The axis of rotary motion of a machine tool member or slide about the X axis.

absolute dimension A dimension expressed with respect to the initial zero point of a coordinate axis.

absolute readout A display of the true slide position as derived from the position commands within the control system.

absolute system NC system in which all positional dimensions, both input and feedback, are measured from a fixed point of origin.

acceleration and deceleration (ACCENDEC) Acceleration and deceleration in feed rate, providing smooth starts and stops when a machine is operating under NC and changes from one feed rate value to another.

access time The time interval between the instant at which information is: **1.** Called for from storage and the instant at which delivery is completed, i.e., the read time. **2.** Ready for storage and the instant at which storage is completed, i.e., the write time.

accuracy **1.** Measured by the difference between the actual position of the machine slide and the position demanded. **2.** Conformity of an indicated value to a true value, i.e., an actual or an accepted standard value. The accuracy of a control system is expressed as the deviation or difference between the ultimately controlled variable and its ideal value, usually in the steady state or at sampled instants.

active storage That part of the control logic which holds the information while it is being transformed into motion.

ADAPT An Air Force adaptation of APT, which has limited vocabulary and can be employed on some small- to medium-sized U.S. computers for NC programming.

adaptive control A technique of automatically adjusting feeds and/or speeds to an optimum by sensing cutting conditions and acting upon them.

ALGOL (see algorithmic language)

algorithm A rule or procedure for solving a mathematical problem that frequently involves repetition of an operation.

algorithmic language (ALGOL) Language used to develop computer programs by algorithm.

alphanumeric or alphameric A system in which the characters used are letters A through Z and numerals 0 to 9.

American Standard Code for Information Interchange (ASCII) A data transmission code which has been established as a standard by the American Standards Association. It is a code in which 7 bits are used to represent each character. (Also known as USACII.)

analog   In NC the term applies to a system which utilizes electrical voltage magnitudes or ratios to represent physical axis positions.

analog-to-digital (A/D) converter   A device that changes physical motion or electrical voltage into digital factors.

APT   (see automatically programmed tools)

arc clockwise   An arc generated by the coordinated motion of two axes in which curvature of the path of the tool with respect to the workpiece is clockwise when the plane of motion is viewed from the positive direction of the perpendicular axis.

arc counterclockwise   Same as arc clockwise above, except substitute "counterclockwise" for "clockwise."

ASCII   (see American Standard Code for Information Interchange)

automatically programmed tools (APT)   A universal computer-assisted program system for multiaxis contouring programming APT III. Provides for five axes of machine tool motion.

automatic system for positioning of tools (AUTOSPOT)   A computer-assigned program for NC positioning and straight-cut systems, developed in the United States by the IBM Space Guidance Center. It is maintained and taught by IBM.

automation   The technique of making a process or system automatic. Automatically controlled operation of an apparatus, process, or system, especially by electronic devices. In present-day terminology, the term is usually used in relation to a system whereby the electronic device controlling an apparatus or process also is interfaced to and communicates with a computer.

AUTOSPOT   (see automatic system for positioning of tools)

auxiliary function   A function of a machine other than the control of the coordinates of a workpiece or cutter—usually ON/OFF type operations.

axis   **1.** A principal direction along which a movement of the tool or workpiece occurs. **2.** One of the reference lines of a coordinate system.

axis inhibit   Prevents movement of the selected slides with the power on.

axis interchange   The capability of inputting the information concerning one axis into the storage of another axis.

axis inversion   The reversal of normal plus-and-minus values along an axis, which makes possible the machining of a left-handed part from right-handed programming or vice versa. Also known as mirror image.

B axis   The axis of rotary motion of a machine tool member or slide about the Y axis.

backlash   A relative movement between interacting mechanical parts, resulting from looseness.

batch processing   A manufacturing operation in which a specified quantity of material is subject to a series of treatment steps. Also, a mode of computer operations in which each program is completed before the next is started.

baud   A unit of signalling speed equal to the number of discrete conditions or signal events per second: 1 bit/s in a train of binary signals and 3 bits/s in an octal train of signals.

BCD (see binary-coded decimal)

binary A numbering system based on 2. Only the digits 0 and 1 are used when written.

binary-coded decimal (BCD) A number code in which individual decimal digits are each represented by a group of binary digits; in the 8-4-2-1 BCD notation, each decimal digit is represented by a four-place binary number, weighted in sequence as 8, 4, 2, and 1.

binary digit (bit) A character used to represent one of the two digits in the binary number system, and the basic unit of information or data storage in a two-state device.

block A set of words, characters, digits, or other elements handled as a unit. On a punched tape, it consists of one or more characters or rows across the tape that collectively provide enough information for an operation. A "word" or group of words considered as a unit separated from other such units by an "end of block" (EOB) character.

block delete Permits selected blocks of tape to be ignored by the control system at the discretion of the operator with permission of the programmer.

boolean algebra An algebra named for George Boole. This algebra is similar in form to ordinary algebra, but with classes, propositions, yes/no criteria, etc., for variables rather than numeric quantities. It includes the operator's AND, OR, NOT, EXCEPT, and IF THEN.

buffer storage 1. A place for storing information in a control for anticipated transference to active storage. It enables a control system to act immediately on stored information without waiting for a tape reader. 2. A register used for intermediate storage of information in the transfer sequence between the computer's accumulators and a peripheral device.

byte A sequence of adjacent bits, usually less than a word, operated on as a unit.

C axis The axis of rotary motion of a machine tool member or slide about the Z axis.

CAD (see computer-aided design)

CAM (see computer-assisted manufacturing)

cancel A command which will discontinue any canned cycle or sequence commands.

canned cycle A preset sequence of events initiated by a single NC command: for example, G84 for NC tap cycle. Also known as fixed cycle.

Cartesian coordinates Means whereby the position of a point can be defined with reference to a set of axes at right angles to each other.

cathode-ray tube (CRT) A display device in which controlled electron beams are used to present alphanumeric or graphical data on a luminescent screen.

central processing unit (CPU) The portion of a computer system consisting of the arithmetic and control units and the working memory.

channel A communication path.

character One of a set of symbols. The general term to include all symbols such as alphabetic letters, numerals, punctuation marks, mathematic operators, etc. Also, the coded representation of such symbols.

chip   A single piece of silicon which has been cut from a slice by scribing and breaking. It can contain one or more circuits but is packaged as a unit.

circular interpolation   **1.** Capability of generating up to 90° of arc using only one block of information as defined by the EIA. **2.** A mode of contouring control which uses the information contained in a single block to produce an arc of a circle.

clear   To erase the contents of a storage device by replacing the contents with blanks or zeros.

clock   A device which generates periodic synchronization signals.

closed loop   A signal path in which outputs are fed back for comparison with desired values to regulate system behavior.

computer   A device capable of accepting information in the form of signals or symbols, performing prescribed operations on the information, and providing results as outputs.

computer-aided design (CAD)   A process which uses a computer to assist in the creation or modification of a design.

computer-assisted manufacturing (CAM)   A process using computer technology to manage and control the operations of a manufacturing facility.

computer numerical control (CNC)   An NC system wherein a dedicated, stored program computer is used to perform some or all of the basic NC functions.

constant surface speed (CSS)   The ability of a turning center MCU to maintain a constant surface speed at the cutting tool point regardless of the changes in work diameter.

continuous path operation   An operation in which rate and direction of relative movement of machine members is under continuous numerical control. There is no pause for data reading.

contouring   An operation in which simultaneous control of more than one axis is accomplished.

contouring control system   An NC system for controlling a machine (milling, drafting, etc.) in a path resulting from the coordinated simultaneous motion of two or more axes.

coordinate dimensioning   A system of dimensioning based on a common starting point.

core memory   A high-speed random-access data storage device utilizing arrays of magnetic ferrite cores, usually employed as a working computer memory.

CPU   (see central processing unit)

CRT   (see cathode-ray tube)

CSS   (see constant surface speed)

cursor   A visual movable pointer used on a CRT by an operator to indicate where corrections or additions are to be made.

cutter diameter compensation   A system in which the programmed path may be altered to allow for the difference between actual and programmed cutter diameters.

cutter offset   **1.** The distance from the part surface to the axial center of a cutter. **2.** An NC feature which allows an operator to use an oversized or undersized cutter.

cutter path   The path described by the center of a cutter.

cycle   **1.** A sequence of operations that is repeated regularly. **2.** The time it takes for one such sequence to occur.

cycle time The period required for a complete action. In particular, the interval required for a read and a write operation in working memory, usually taken as a measure of computer speed.

debug To detect, locate, and remove mistakes from computer software or hardware.

decimal code A code in which each allowable position has 1 of 10 possible states. (The conventional decimal-number system is a decimal code.)

decoder A circuit arrangement which receives and converts digital information from one form to another.

digital Representation of data in discrete or numerical form.

digital computer A computer that operates on symbols representing data by performing arithmetic and logic operations.

digital-to-analog (D-A) conversion Production of an analog signal whose instantaneous magnitude is proportional to the value of a digital input.

digitizer A device which tracks the relative position of a cursor, for the purpose of recording relative locations of a part print or actual part.

direct numerical control (DNC) NC of machining or processing by a computer.

disk memory A nonprogrammable, bulk-storage, random-access memory consisting of a magnetizable coating on one or both sides of a rotating thin circular plate.

display Lights, annunciators, numerical indicators, or other operator output devices at consoles or remote stations.

DNC (see direct numerical control)

documentation The group of techniques necessarily used to organize, present, and communicate recorded specialized knowledge.

downtime The interval during which a device is inoperative.

dump To copy the present contents of a memory onto a printout or auxiliary storage.

dwell A timed delay of programmed or established duration, not cyclic or sequential, i.e., not an interlock or hold.

edit To modify a program, or alter stored data prior to output.

Electronics Industries Association (EIA) standard code Standard codes for positioning, straight-cut, and contouring control systems.

encoder An electromechanical transducer which produces a serial or parallel digital indication of mechanical angle or displacement.

end-of-block (EOB) character 1. A character indicating the end of a block of tape information. Used to stop the tape reader after a block has been read. 2. The typewriter function of the carriage return when machine control tapes are prepared.

end of program A miscellaneous function (M02) indicating completion of a workpiece. (Stops spindle, coolant, and feed after completion of all commands in the block. Used to reset control and/or machine.)

end of tape A miscellaneous function (M30) which stops spindle, coolant, and feed after completion of all commands in the block. (Used to reset control and/or machine.)

EOB character (see end-of-block character)

error signal Difference between the output and input signals in a servo system.

executive Software which controls the execution of programs in the computer, based on established priorities and real-time or demand requirements.

feedback The signal or data fed back to a commanding unit from a controlled machine or process to denote its response to the command signal; the signal represents the difference between actual response and desired response that is used by the commanding unit to improve performance of the controlled machine or process.

feedback device An element of a control system which converts linear or rotary motion to an electrical signal for comparison to the input signal, e.g., resolver, encoder.

feed engage point The point where the motion of the Z axis changes from rapid traverse to a programmed feed (usually referred to as the R dimension).

feed function The relative motion between the tool or instrument and the work due to motion of the programmed axis or axes.

feed override A variable manual control function directing the control system to reduce or increase the programmed feedrate.

fixed block format A format in which the number and sequence of words and characters appearing in successive blocks is constant.

fixed cycle (see canned cycle)

fixed sequence format A means of identifying a word by its location in a block of information. Words must be presented in a specific order, and all possible words preceding the last desired word must be present in the block.

floating zero A characteristic of an MCU which allows the zero reference point on an axis to be set at any point of the machine table travel.

format classification A means, usually in an abbreviated notation, by which the motions, dimensional data, type of control system, number of digits, auxiliary functions, etc., for a particular system can be denoted.

formula translator (FORTRAN) An algebraic procedure-oriented computer language designed to solve arithmetic and logical programs.

full range floating zero A characteristic of a numerical machine tool control permitting the zero point on an axis to be readily shifted over a specified range. The control retains information on the location of ''permanent'' zero.

G code A word addressed by the letter G and followed by a numerical code, defining preparatory functions or cycle types in an NC system.

gage height A predetermined partial retraction point along the Z axis to which the cutter retreats from time to time to allow safe XY table travel.

hard copy Any form of computer-produced printed document. Also, sometimes punch cards or paper tape.

hardware Physical equipment.

head A device, usually a small electromagnet on a storage medium such as magnetic tape or a magnetic drum, that reads, records, or erases information on that medium. The block assembly and perforating or reading fingers used for punching or reading holes in paper tape.

Hollerith   A 12-bit code used for recording characters in punched paper cards.

IC   (see integrated circuit)

incremental dimension   A dimension expressed with respect to the preceding point in a sequence of points.

incremental system   Control system in which each coordinate or positional dimension is taken from the last position.

indexing   Movement of one axis at a time to a precise point from numeric commands.

inhibit   To prevent an action or acceptance of data by applying an appropriate signal to the appropriate input.

input   A dependent variable applied to a control unit or system.

instruction   A statement that specifies an operation and the values or locations of its operands.

integrated circuit (IC)   A combination of interconnected passive and active circuit elements incorporated on a continuous substrate.

interface   1. A hardware component or circuit linking two pieces of electrical equipment having separate functions, e.g., tape reader to data processor or control system to machine. 2. A hardware component or circuit for linking the computer to an external input/output (I/O) device.

interpolation   1. The insertion of intermediate information based on assumed order or computation. 2. A function of a control whereby data points are generated between given coordinate positions to allow simultaneous movement of two

or more axes of motion in a defined geometric pattern, e.g., linear, circular, and parabolic.

interpolator   A device which is part of an NC system and performs interpolation.

jog   A control function which provides for the momentary operation of a drive for the purpose of accomplishing a small movement of the driven machine.

large-scale integration (LSI)   A large number of interconnected ICs manufactured simultaneously on a single slice of semiconductor material (usually over 100 gates or basic circuits, with at least 500 circuit elements).

leading zeros   Redundant zeros to the left of a decimal point or number.

linear interpolation   A function of a control whereby data points are generated between given coordinate positions to allow simultaneous movement of two or more axes of motion in a linear (straight-line) path.

logic   1. Electronic devices used to govern a particular sequence of operations in a given system. 2. Interrelation or sequence of facts or events when seen as inevitable or predictable.

LSI   (see large-scale integration)

machining center   A machine tool, usually numerically controlled, capable of automatically drilling, reaming, tapping, milling, and boring multiple faces of a part and often equipped with a system for automatically changing cutting tools.

macro   A source language instruction from which many machine language instructions can be generated.

magnetic core   An element for switching or

storing a bit of information in a computer. Thousands of cores may be used in one computer.

management information system (MIS) An information feedback system from the machine to management and implemented by a computer.

manual data input (MDI) A means of inserting data manually into the control system.

manual part programming The manual preparation of a manuscript in machine control language and format to define a sequence of commands for use on an NC machine.

manuscript Form used by a part programmer for listing detailed manual or computer part programming instructions.

MDI (see manual data input)

medium-scale integration (MSI) Smaller than LSI, but having at least 12 gates or basic circuits with at least 100 circuit elements (see large-scale integration).

memory A device or medium used to store information in a form that can be understood by the computer hardware.

microprogramming A programming technique in which multiple-instruction operations can be combined for greater speed and more efficient memory use.

microsecond One millionth of a second.

millisecond One thousandth of a second.

mirror image (see axis inversion)

MIS (see management information system)

modal A set of commands retained in a system until a new command cancels or replaces them.

nanosecond A billionth of a second; a common unit of measure of computer operating speed.

numerical control (NC) A technique of operating machine tools or similar equipment in which motion is developed in response to numerically coded commands.

off-line Operating software or hardware not under the direct control of a central processor, or operations performed while a computer is not monitoring or controlling processes or equipment.

offset The steady-state deviation of the controlled variable from a fixed setpoint.

on-line A condition in which equipment or programs are under direct control of a central processor.

open-loop system A control system that has no means of comparing the output with the input for control purposes (no feedback).

optional stop A Miscellaneous Function command similar to Program Stop except that the control ignores the command unless the operator has previously pushed a button to validate the command (M01).

output Dependent variable signal produced by a transmitter, control unit, or other device.

overshoot The amount that a controlled variable exceeds its desired value after a change of input.

parabola A plane curve generated by a point moving so that its distance from a fixed second point is equal to its distance from a fixed line.

parabolic interpolation Control of cutter path by interpolation between three fixed points by assuming the intermediate points are on a parabola.

parity check　A test of whether the number of 1s or 0s in an array of binary digits is odd or even to detect errors in a group of bits.

part program　Specific and complete set of data and instructions written in source languages for computer processing or written in machine language for manual programming for the purpose of manufacturing a part of an NC machine.

part programmer　A person who prepares the planned sequence of events for the operation of an NC machine tool.

PAU　(see position analog unit)

picosecond　One millionth of one microsecond.

point-to-point control system　An NC system which controls motion only to reach a given end point but exercises no path control during the transition from one end point to the next.

polar coordinates　A mathematical system for locating a point in a plane by the length of its radius vector and the angle this vector makes with a fixed line.

position analog unit (PAU)　The unit which feeds back, to the servo amplifier, analog information corresponding to the position of the machine slide to be compared with positional input information.

positioning/contouring　A type of NC system that has the capability of contouring, without buffer storage, in two axes, and the ability of positioning in a third axis for such operations as drilling, tapping, boring, etc.

postprocessor　The part of the software which converts all the cutter path coordinate data (obtained from the general-purpose processor and all other programming instructions and specifications for the particular machine and control) into a form which the machine control can interpret correctly.

preparatory function　An NC command on the input tape changing the mode of operation of the control (generally noted at the beginning of a block by "G" plus two digits).

printed circuit　A circuit for electronic components made by depositing conductive material in continuous paths from terminal to terminal on an insulating surface.

program　1. A plan for the solution of a problem. A complete program includes plans for the transcription of data, coding for the computer, and absorption of the results into the system. The list of coded instructions is called a *routine*. 2. To plan a computation or process from the asking of a question to the delivery of the results, including the integration of the operation into an existing system. Thus, programming consists of planning and coding, including numerical analysis, systems analysis, specification of printing formats, and any other functions necessary to the integration of a computer in a system.

programmed dwell　The capability of commanding delays in program execution for a programmable length of time.

program stop　A Miscellaneous Function (M00) command to stop the spindle, coolant, and feed after completion of the dimensional move commanded in the block. To continue with the remainder of the program, the operator must initiate a restart.

punch card　A piece of lightweight cardboard on which information is represented by holes punched in specific positions.

punched paper tape A strip of paper on which characters are represented by combinations of holes.

pulse A short-duration change in the level of a variable.

quadrant Any of the four parts into which a plane is divided by rectangular coordinate axes lying in that plane.

R dimension (see feed engage point)

random-access memory (RAM) A form of temporary internal storage contents which can be recalled and changed by the user; also known as read-and-write memory.

reader A device capable of sensing information stored in off-line memory media (cards, paper tape, magnetic tape, etc.) and generating equivalent information in an on-line memory device (e.g., register or memory locations).

read-only memory (ROM) Permanent internal memory containing data or operating instructions that can be recalled but not changed by the user.

real-time clock The circuitry which maintains time for use in program execution and event initiation.

reference block A block within an NC program identified by an O or H in place of the word address N and containing sufficient data to enable resumption of the program following an interruption. (This block should be located at a convenient point in the program to enable the operator to reset and resume operation.)

resolver 1. A mechanical-to-electrical transducer whose input is a vector quantity and whose outputs are components of the vector. 2. A transformer whose coupling may be varied by rotating one set of windings relative to another. It consists of a stator and rotor, each having two distributed windings 90 electrical degrees apart.

ROM (see read-only memory)

routine A series of computer instructions which performs a specific task.

scale To change a quantity by a given factor, to bring its range within prescribed limits.

scanner The equipment used to digitize coordinate information from a master and convert it to punched tape for later re-creation of the master shape on an NC machine.

sequence number A number identifying the relative locations of blocks or groups of blocks on a tape.

servo amplifier The part of the servo system which increases the error signal and provides the power to drive the machine slides or the servo valve controlling a hydraulic drive.

setpoint The position established by an operator as the starting point for the program on an NC machine.

significant digit A digit that contributes to the precision of a numeral. The number of significant digits is counted beginning with the digit contributing the most value, called the most significant digit, and ending with the one contributing the least value, called the least significant digit.

software The collection of programs, routines, and documents associated with a computer.

storage A memory device in which data can be entered and held and from which it can be retrieved.

subroutine A series of computer instructions to perform a specific task for many other routines. It is distinguishable from a main routine in that it requires, as one of its parameters, a location specifying where to return to the main program after its function has been accomplished.

tape A magnetic or perforated paper medium for storing information.

tool function A tape command identifying a tool and calling for its selection. The address is normally a T word.

tool length compensation A manual input means which eliminates the need for preset tooling and allows the programmer to program all tools as if they were of equal length.

tool offset 1. A correction for tool position parallel to a controlled axis. 2. The ability to reset tool position manually to compensate for tool wear, finish cuts, and tool exchange.

tool path The center line of an NC cutting tool while a cutting operation such as milling, drilling, or boring is performed.

trailing zero suppression Same as *zero suppression.*

turning center A lathe-type NC machine tool capable of automatically boring, turning outer and inner diameters, threading, and facing multiple diameters and faces of a part, and often equipped with a system for automatically changing or indexing cutting tools.

turnkey system A term applied to an agreement whereby a supplier will install an NC or computer system and have total responsibility for building, installing, and testing the system.

USACII (see American Standard Code for Information Interchange)

variable block format Tape format which allows the number of words in successive blocks to vary.

vector A quantity that has magnitude, direction, and sense, and that is commonly represented by a directed line segment whose length represents the magnitude and whose orientation in space represents the direction.

vector feed rate The resultant feed rate at which a cutter or tool moves with respect to the work surface. The individual slides may move slower or faster than the programmed rate, but the resultant movement is equal to the programmed rate.

word address format Addressing each word in a block by one or more characters which identify the meaning of the word.

word length The number of bits or characters in a word.

X axis Axis of motion that is always horizontal and parallel to the workholding surface.

Y axis Axis of motion that is perpendicular to both X and Z axes.

Z axis Axis of motion that is always parallel to the principal spindle of the machine.

zero offset A characteristic of a numerical machine tool control permitting the zero point on an axis to be shifted readily over a specified range. The control retains information on the location of the "permanent" zero.

zero shift Same as zero offset, except the control does *not* retain information on the location of the "permanent" zero.

zero suppression The elimination of nonsignificant zeros to the left of significant digits, usually before printing.

APPENDIX

Table 1　CONVERSION OF METRIC AND ENGLISH MEASURES

	Conversion of inches to millimeters						Conversion of millimeters to inches				
Inches	Milli-meters	Inches	Milli-meters	Inches	Milli-meters	Milli-meters	Inches	Milli-meters	Inches	Milli-meters	Inches
.001	0.025	.290	7.37	.660	16.76	0.01	.0004	0.35	.0138	0.68	.0268
.002	0.051	.300	7.62	.670	17.02	0.02	.0008	0.36	.0142	0.69	.0272
.003	0.076	.310	7.87	.680	17.27	0.03	.0012	0.37	.0146	0.70	.0276
.004	0.102	.320	8.13	.690	17.53	0.04	.0016	0.38	.0150	0.71	.0280
.005	0.127	.330	8.38	.700	17.78	0.05	.0020	0.39	.0154	0.72	.0283
.006	0.152	.340	8.64	.710	18.03	0.06	.0024	0.40	.0157	0.73	.0287
.007	0.178	.350	8.89	.720	18.29	0.07	.0028	0.41	.0161	0.74	.0291
.008	0.203	.360	9.14	.730	18.54	0.08	.0031	0.42	.0165	0.75	.0295
.009	0.229	.370	9.40	.740	18.80	0.09	.0035	0.43	.0169	0.76	.0299
.010	0.254	.380	9.65	.750	19.05	0.10	.0039	0.44	.0173	.077	.0303
.020	0.508	.390	9.91	.760	19.30	0.11	.0043	0.45	.0177	0.78	.0307
.030	0.762	.400	10.16	.770	19.56	0.12	.0047	0.46	.0181	0.79	.0311
.040	1.016	.410	10.41	.780	19.81	0.13	.0051	0.47	.0185	0.80	.0315
.050	1.270	.420	10.67	.790	20.07	0.14	.0055	.048	.0189	0.81	.0319
.060	1.524	.430	10.92	.800	20.32	0.15	.0059	0.49	.0193	0.82	.0323
.070	1.778	.440	11.18	.810	20.57	0.16	.0063	0.50	.0197	0.83	.0327
.080	2.032	.450	11.43	.820	20.83	0.17	.0067	0.51	.0201	0.84	.0331
.090	2.286	.460	11.68	.830	21.08	0.18	.0071	0.52	.0205	0.85	.0335
.100	2.540	.470	11.94	.840	21.34	0.19	.0075	0.53	.0209	0.86	.0339
.110	2.794	.480	12.19	.850	21.59	0.20	.0079	0.54	.0213	0.87	.0343
.120	3.048	.490	12.45	.860	21.84	0.21	.0083	0.55	.0217	0.88	.0346
.130	3.302	.500	12.70	.870	22.10	0.22	.0087	0.56	.0220	0.89	.0350
.140	3.56	.510	12.95	.880	22.35	0.23	.0091	0.57	.0224	0.90	.0354
.150	3.81	.520	13.21	.890	22.61	0.24	.0094	0.58	.0228	0.91	.0358
.160	4.06	.530	13.46	.900	22.86	0.25	.0098	0.59	.0232	0.92	.0362
.170	4.32	.540	13.72	.910	23.11	0.26	.0102	0.60	.0236	0.93	.0366
.180	4.57	.550	13.97	.920	23.37	0.27	.0106	0.61	.0240	0.94	.0370
.190	4.83	.560	14.22	.930	23.62	0.28	.0110	0.62	.0244	0.95	.0374
.200	5.08	.570	14.48	.940	23.88	0.29	.0114	0.63	.0248	0.96	.0378
.210	5.33	.580	14.73	.950	24.13	0.30	.0118	0.64	.0252	0.97	.0382
.220	5.59	.590	14.99	.960	24.38	0.31	.0122	0.65	.0256	0.98	.0386
.230	5.84	.600	15.24	.970	24.64	0.32	.0126	0.66	.0260	0.99	.0390
.240	6.10	.610	15.49	.980	24.89	0.33	.0130	0.67	.0264	1.00	.0394
.250	6.35	.620	15.75	.990	25.15	0.34	.0134	....	....	....	....
.260	6.60	.630	16.00	1.000	25.40						
.270	6.86	.640	16.26	....	....						
.280	7.11	.650	16.51	....	....						

*Courtesy Automatic Electric Company.

Table 2   VALUES OF FRACTIONAL SIZES EXPRESSED IN MILLIMETERS*†

25.4 mm equals 1 inch

Fractional sizes		1 in.	2 in.	3 in.	4 in.	5 in.	6 in.	Fractional sizes		1 in.	2 in.	3 in.	4 in.	5 in.	6 in.
		25.4	50.8	76.2	101.6	127.	152.4	1/2	12.7	38.1	63.5	88.9	114.3	139.7	165.1
1/64	0.40	25.80	51.20	76.60	102.	127.39	152.79	33/64	13.10	38.49	63.90	89.3	114.69	140.09	165.49
1/32	0.79	26.19	51.59	76.99	102.39	127.79	153.19	17/32	13.49	38.89	64.29	89.69	115.09	140.49	165.89
3/64	1.19	26.59	51.99	77.39	102.79	128.19	153.59	35/64	13.89	39.29	64.69	90.09	115.49	140.89	166.29
1/16	1.59	26.99	52.39	77.79	103.19	128.59	153.98	9/16	14.29	39.69	65.09	90.49	115.89	141.29	166.68
5/64	1.98	27.38	52.78	78.18	103.58	128.98	154.38	37/64	14.68	40.08	65.48	90.88	116.28	141.68	167.08
3/32	2.38	27.78	53.18	78.58	103.98	129.38	154.78	19/32	15.08	40.48	65.88	91.28	116.68	142.08	167.48
7/64	2.77	28.17	53.58	78.98	104.37	129.78	155.18	39/64	15.48	40.88	66.28	91.68	117.08	142.48	167.88
1/8	3.17	28.57	53.97	79.37	104.77	130.17	155.57	5/8	15.87	41.27	66.67	92.07	117.47	142.87	168.27
9/64	3.57	28.97	54.37	79.77	105.17	130.57	155.97	41/64	16.27	41.67	67.07	92.47	117.87	143.27	168.67
5/32	3.97	29.37	54.77	80.17	105.57	130.97	156.37	21/32	16.67	42.07	67.47	92.87	118.27	143.67	169.07
11/64	4.37	29.76	55.16	80.56	105.96	131.36	156.76	43/64	17.07	42.46	67.86	93.26	118.66	144.06	169.46
3/16	4.76	30.16	55.56	80.96	106.36	131.76	157.16	11/16	17.46	42.86	68.26	93.66	119.06	144.46	169.86
13/64	5.16	30.56	55.96	81.36	106.76	132.16	157.56	45/64	17.86	43.26	68.66	94.06	119.46	144.86	170.26
7/32	5.56	30.96	56.36	81.75	107.16	132.55	157.95	23/32	18.26	43.66	69.05	94.45	119.85	145.25	170.65
15/64	5.95	31.35	56.75	82.15	107.55	132.95	158.35	47/64	18.65	44.05	69.45	94.85	120.25	145.65	171.05
1/4	6.35	31.75	57.15	82.55	107.95	133.35	158.75	3/4	19.05	44.45	69.85	95.25	120.65	146.05	171.45
17/64	6.75	32.15	57.55	82.95	108.34	133.74	159.14	49/64	19.45	44.85	70.25	95.65	121.04	146.44	171.84
9/32	7.14	32.54	57.94	83.34	108.74	134.14	159.54	25/32	19.84	45.24	70.64	96.04	121.44	146.84	172.24
9/64	7.54	32.94	58.34	83.74	109.14	134.54	159.94	51/64	20.24	45.64	71.04	96.44	121.84	147.24	172.64
5/16	7.94	33.34	58.74	84.14	109.54	134.94	160.33	13/16	20.64	46.04	71.44	96.84	122.24	147.63	173.03
21/64	8.33	33.73	59.13	84.53	109.93	135.33	160.73	53/64	21.03	46.43	71.83	97.23	122.63	148.03	173.43
11/32	8.73	34.13	59.53	84.93	110.33	135.73	161.13	27/32	21.43	46.83	72.23	97.63	123.03	148.43	173.83
23/64	9.13	34.53	59.93	85.33	110.73	136.13	161.53	55/64	21.83	47.23	72.63	98.03	123.43	148.83	174.22
3/8	9.52	34.92	60.32	85.72	111.12	136.52	161.92	7/8	22.22	47.62	73.02	98.42	123.82	149.22	174.62
25/64	9.92	35.32	60.72	86.12	111.52	136.92	162.32	57/64	22.62	48.02	73.42	98.82	124.22	149.62	175.02
13/32	10.32	35.72	61.12	86.52	111.92	137.32	162.72	29/32	23.02	48.42	73.82	99.22	124.62	150.02	175.42
27/64	10.72	36.11	61.51	86.91	112.31	137.71	163.11	59/64	23.42	48.81	74.21	99.61	125.01	150.41	175.81
7/16	11.11	36.51	61.91	87.31	112.71	138.11	163.51	15/16	23.81	49.21	74.61	100.01	125.41	150.81	176.21
29/64	11.51	36.91	62.31	87.71	113.11	138.51	163.91	61/64	24.21	49.61	75.01	100.41	125.81	151.21	176.61
15/32	11.91	37.31	62.71	88.1	113.5	138.9	164.3	31/32	24.61	50.01	75.4	100.8	126.2	151.6	177.
31/64	12.3	37.7	63.1	88.5	113.9	139.3	164.7	63/64	25.	50.4	75.8	101.2	126.6	152.	177.4

*To use, read down appropriate inch column to the desired fraction line. The number indicated is the size in millimeters.
†Courtesy The Cleveland Twist Drill Co.

| Table 3 | | **METRIC-ENGLISH CONVERSION TABLE** |

Multiply	By	To get equivalent number of	Multiply	By	To get equivalent number of
	Length			**Acceleration**	
Inch	25.4	Millimeters (mm)	Foot/second^2	0.304 8	Meter per second^2 (m/s^2)
Foot	0.304 8	Meters (m)	Inch/second^2	0.025 4	Meter per second^2
Yard	0.914 4	Meters			
Mile	1.609	Kilometers (km)		**Torque**	
	Area		Pound-inch	0.112 98	Newton-meters (N-m)
			Pound-foot	1.355 8	Newton-meters
Inch^2	645.2	Millimeters^2 (mm^2)			
	6.45	Centimeters^2 (cm^2)		**Power**	
Foot^2	0.092 9	Meters^2 (m^2)			
Yard^2	0.836 1	Meters^2	Horsepower	0.746	Kilowatts (kW)
	Volume			**Pressure or stress**	
Inch^3	16 387.	mm^3	Inches of water	0.249 1	Kilopascals (kPa)
	16.387	cm^3	Pounds/square inch	6.895	Kilopascals
	0.016 4	Liters (l)			
Quart (U.S.)	0.946 4	Liters		**Energy or work**	
Quart (imperial)	1.136	Liters	BTU	1 055.	Joules (J)
Gallon (U.S.)	3.785 4	Liters	Foot-pound	1.355 8	Joules
Gallon (imperial)	4.459	Liters	Kilowatthour	3 600 000.	Joules (J = one W's)
Yard^3	0.764 6	Meters^3 (m^3)		or 3.6 × 10^6	
	Mass			**Light**	
Pound	0.453 6	Kilograms (kg)	Footcandle	1.076 4	Lumens per meter^2 (lm/m^2)
Ton	907.18	Kilograms (kg)			
Ton	0.907	Tonne (t)		**Fuel performance**	
	Force		Miles per gallon	0.425 1	Kilometers per liter (km/l)
			Gallons per mile	2.352 7	Liters per kilometers (l/km)
Kilogram	9.807	Newtons (N)			
Ounce	0.278 0	Newtons		**Velocity**	
Pound	4.448	Newtons			
	Temperature		Miles per hour	1.609 3	Kilometers per hr (km/h)
Degree Fahrenheit	(°F − 32) ÷ 1.8	Degree Celsius (C)			
Degree Celsius	(°C × 1.8) + 32	Degree Fahrenheit (F)			

Table 4 FORMULA SHORTCUTS

For the correct formula, block out (cover) the unknown; the remainder is the formula. In each diagram the horizontal line is the division line; the vertical line(s) is the multiplication line.

Code:
A = Area
C = Circumference
CS = Cutting speed
D = Diameter

L = Length
R = Radius
r/min = Revolutions/minute
S = Strokes/minute

b = Base
h = Height
m = Meters
mm = Millimeters

Circle

$C = \pi \times D$

Division line →

$D = \dfrac{C}{\pi}$

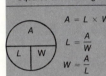

↑ Multiplication line

Four-element formulas
1. Block out unknown.
2. Cross-multiply diagonally opposite elements.
3. Divide by remaining element.

Triangles	Shaper speed

Triangles

$A = \dfrac{b \times h}{2}$

$b = \dfrac{A \times 2}{h}$

$h = \dfrac{A \times 2}{b}$

Shaper speed

$CS = \dfrac{S \times L}{7}$

$S = \dfrac{CS \times 7}{L}$

$L = \dfrac{CS \times 7}{S}$

Area

Revolutions per minute (r/min)
(Lathe, drill, mill, grinder)

Squares and rectangles	Circles	Inch	Metric

Squares and rectangles

$A = L \times W$

$L = \dfrac{A}{W}$

$W = \dfrac{A}{L}$

Circles

$A = \pi \times R^2$

$R^2 = \dfrac{A}{\pi}$

Inch

$r/min = \dfrac{CS\ (ft) \times 4}{D\ (in.)}$

$CS = \dfrac{r/min \times D}{4}$

$D = \dfrac{CS \times 4}{r/min}$

Metric

$r/min = \dfrac{CS\ (m) \times 320}{D\ (mm)}$

$CS = \dfrac{r/min \times D}{320}$

$D = \dfrac{CS \times 320}{r/min}$

Table 5 ISO METRIC PITCH AND DIAMETER COMBINATIONS

Nominal diameter, mm	Thread pitch, mm	Nominal diameter, mm	Thread pitch, mm
1.6	0.35	20	2.5
2	0.40	24	3.0
2.5	0.45	30	3.5
3	0.50	36	4.0
3.5	0.60	42	4.5
4	0.70	48	5.0
5	0.80	56	5.5
6.0	1.00	64	6.0
8	1.25	72	6.0
10	1.50	80	6.0
12	1.75	90	6.0
14	2.00	100	6.0
16	2.00		

Table 6 REVOLUTIONS PER MINUTE FOR WORK OR CUTTING TOOL DIAMETERS

Diameter		Cutting speed, ft/min or m/min							
Inches	mm	40 ft or 12 m	60 ft or 18 m	80 ft or 24 m	90 ft or 27 m	100 ft or 30 m	120 ft or 36 m	150 ft or 45 m	200 ft or 60 m
1/8	3	1280	1920	2560	2880	3200	3840	4800	6400
1/4	6	640	960	1280	1440	1600	1920	2400	3200
3/8	9	427	640	853	960	1067	1280	1600	2133
1/2	13	320	480	640	720	800	960	1200	1600
5/8	16	256	384	512	576	640	768	960	1280
3/4	19	213	320	427	480	533	640	800	1067
7/8	22	183	274	366	411	457	548	686	914
1	25	160	240	320	360	400	480	600	800
1 1/4	32	128	192	256	288	320	384	480	640
1 1/2	38	107	160	213	240	267	320	400	533
1 3/4	44	91	137	183	206	229	274	343	457
2	50	80	120	160	180	200	240	300	400
2 1/2	63	64	96	128	144	160	192	240	320
3	76	53	80	107	120	133	160	200	267
3 1/2	89	46	68	91	103	114	136	171	229
4	100	40	60	80	90	100	120	150	200

Table 7 **SOLUTIONS FOR RIGHT-ANGLE TRIANGLES**

$\text{Sine} \angle = \dfrac{\text{Side opposite}}{\text{Hypotenuse}}$	$\text{Cosecant} \angle = \dfrac{\text{Hypotenuse}}{\text{Side opposite}}$
$\text{Cosine} \angle = \dfrac{\text{Side adjacent}}{\text{Hypotenuse}}$	$\text{Secant} \angle = \dfrac{\text{Hypotenuse}}{\text{Side adjacent}}$
$\text{Tangent} \angle = \dfrac{\text{Side opposite}}{\text{Side adjacent}}$	$\text{Cotangent} \angle = \dfrac{\text{Side adjacent}}{\text{Side opposite}}$

Knowing	Formulas to find	
Sides a & b	$c = \sqrt{a^2 - b^2}$	$\sin B = \dfrac{b}{a}$
Side a & angle B	$b = a \times \sin B$	$c = a \times \cos B$
Sides a & c	$b = \sqrt{a^2 - c^2}$	$\sin C = \dfrac{c}{a}$
Side a & angle C	$b = a \times \cos C$	$c = a \times \sin C$
Sides b & c	$a = \sqrt{b^2 + c^2}$	$\tan B = \dfrac{b}{c}$
Side b & angle B	$a = \dfrac{b}{\sin B}$	$c = b \times \cot B$
Side b & angle C	$a = \dfrac{b}{\cos C}$	$c = b \times \tan C$
Side c & angle B	$a = \dfrac{c}{\cos B}$	$b = c \times \tan B$
Side c & angle C	$a = \dfrac{c}{\sin C}$	$b = c \times \cot C$

Table 8A   COORDINATE FACTORS AND ANGLES 3-HOLE DIVISION

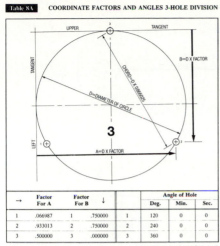

→	Factor For A		Factor For B	↓		Angle of Hole		
						Deg.	Min.	Sec.
1	.066987	1	.750000	1		120	0	0
2	.933013	2	.750000	2		240	0	0
3	.500000	3	.000000	3		360	0	0

Courtesy W. J. Woodworth and J. D. Woodworth

Table 8B   COORDINATE FACTORS AND ANGLES 4-HOLE DIVISION

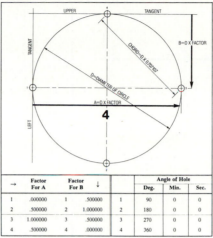

→	Factor For A		Factor For B	↓		Angle of Hole		
						Deg.	Min.	Sec.
1	.000000	1	.500000	1		90	0	0
2	.500000	2	1.000000	2		180	0	0
3	1.000000	3	.500000	3		270	0	0
4	.500000	4	.000000	4		360	0	0

Courtesy W. J. Woodworth and J. D. Woodworth

Table 8C   COORDINATE FACTORS AND ANGLES 5-HOLE DIVISION

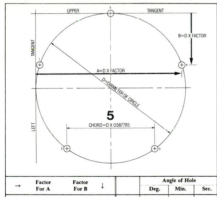

→	Factor For A		Factor For B	↓		Angle of Hole		
						Deg.	Min.	Sec.
1	024472	1	.345492	1		72	0	0
2	.205107	2	.904508	2		144	0	0
3	.793893	3	.904508	3		216	0	0
4	.975528	4	.345492	4		288	0	0
5	.500000	5	.000000	5		360	0	0

Courtesy W. J. Woodworth and J. D. Woodworth

Table 8D   COORDINATE FACTORS AND ANGLES 6-HOLE DIVISION

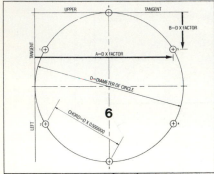

→	Factor For A		Factor For B	↓		Angle of Hole		
						Deg.	Min.	Sec.
1	.066987	1	.250000	1		60	0	0
2	.066987	2	.750000	2		120	0	0
3	.500000	3	1.000000	3		180	0	0
4	.933013	4	.750000	4		240	0	0
5	.933013	5	.250000	5		300	0	0
6	.500000	6	.000000	6		360	0	0

Courtesy W. J. Woodworth and J. D. Woodworth

Table 8E   COORDINATE FACTORS AND ANGLES 7-HOLE DIVISION

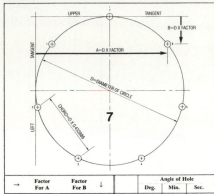

→	Factor For A		Factor For B	↓		Angle of Hole		
						Deg.	Min.	Sec.
1	.109084	1	.188255	1		51	25	42-6 /7
2	.012536	2	.611261	2		102	51	23-5 /7
3	.283058	3	.950484	3		154	17	8-4 /7
4	.716942	4	.950484	4		205	42	51-3 /7
5	.987464	5	.611261	5		257	8	34-2 /7
6	.890916	6	.188255	6		308	34	17-1 /7
7	.500000	7	.000000	7		360	0	0

Courtesy W. J. Woodworth and J. D. Woodworth

Table 8F   COORDINATE FACTORS AND ANGLES 8-HOLE DIVISION

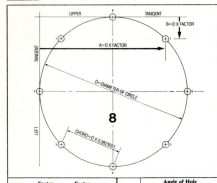

→	Factor For A		Factor For B	↓		Angle of Hole		
						Deg.	Min.	Sec.
1	.146447	1	.146447	1		45	0	0
2	.000000	2	.500000	2		90	0	0
3	.146447	3	.853553	3		135	0	0
4	.500000	4	1.000000	4		180	0	0
5	.853553	5	.853553	5		225	0	0
6	1.000000	6	.500000	6		270	0	0
7	.853553	7	.146447	7		315	0	0
8	.500000	8	.000000	8		360	0	0

Courtesy W. J. Woodworth and J. D. Woodworth

Table 8G COORDINATE FACTORS AND ANGLES 9-HOLE DIVISION

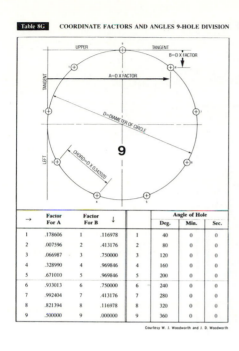

→	Factor For A		Factor For B	↓		Angle of Hole		
						Deg.	Min.	Sec.
1	.178606	1	.116978	1		40	0	0
2	.007596	2	.413176	2		80	0	0
3	.066987	3	.750000	3		120	0	0
4	.328990	4	.969846	4		160	0	0
5	.671010	5	.969846	5		200	0	0
6	.933013	6	.750000	6		240	0	0
7	.992404	7	.413176	7		280	0	0
8	.821394	8	.116978	8		320	0	0
9	.500000	9	.000000	9		360	0	0

Courtesy W. J. Woodworth and J. D. Woodworth

Table 8H COORDINATE FACTORS AND ANGLES 10-HOLE DIVISION

→	Factor For A		Factor For B	↓		Angle of Hole		
						Deg.	Min.	Sec.
1	.206107	1	.095492	1		36	0	0
2	.024472	2	.345492	2		72	0	0
3	.024472	3	.654508	3		108	0	0
4	.206107	4	.904508	4		144	0	0
5	.500000	5	1.000000	5		180	0	0
6	.793893	6	.904508	6		216	0	0
7	.975528	7	.654508	7		252	0	0
8	.975528	8	.345492	8		288	0	0
9	.793893	9	.095492	9		324	0	0
10	.500000	10	.000000	10		360	0	0

Courtesy W. J. Woodworth and J. D. Woodworth

Table 8I COORDINATE FACTORS AND ANGLES 11-HOLE DIVISION

→	Factor For A		Factor For B	↓		Angle of Hole		
						Deg.	Min.	Sec.
1	.229680	1	.079373	1		32	43	38-2/11
2	.045184	2	.292293	2		65	27	16-4/11
3	.005089	3	.571157	3		98	10	54-6/11
4	.122125	4	.827430	4		130	54	32-8/11
5	.359134	5	.979746	5		163	38	10-10/11
6	.640866	6	.979746	6		196	21	49-1/11
7	.877875	7	.827430	7		229	5	27-3/11
8	.994911	8	.571157	8		261	49	5-5/11
9	.954816	9	.292293	9		294	32	43-7/11
10	.770320	10	.079373	10		327	16	21-9/11
11	.500000	11	.000000	11		360	0	0

Courtesy W. J. Woodworth and J. D. Woodworth

Table 9 **AN NC MILLING AND DRILLING EXERCISE.**

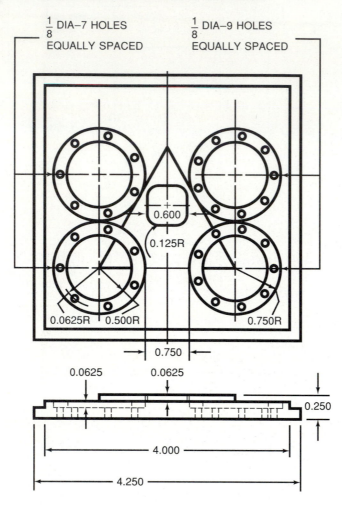

Index

Abacus, 26–27
Abrasives, 209
Absolute programming and positioning, 18–20, 55, 57
  in CNC systems, 84
  MCU register for, 146
  preparatory functions for, 104, 241
Acceleration function, 102
Accuracy:
  improvements in, with NC, 21–22, 69–71
  of servo systems, 78–79
ACTION language, 193–194
ADAPT language, 192
Adaptive control:
  for machining centers, 238–239
  preparatory function for, 103
Addition, binary, 5–7
Address character functions, 94
Address keys for EDM, 307
Advantages of numerical control, 20–22, 67–75, 88–90
AI (artificial intelligence), 30
Air defense, computers for, 32
Alignment of workpiece with EDM, 308
Alphanumeric decoders, 75
American Standard Code for Information Interchange (see ASCII coding system)
Analog computers, 33
Analog feedback systems, 77
Anchor program, 325–327
Angle plates, 224

Angles, basic tool, 204
APT (Automatic Programming of Tools) language, 191–192
  general processor for, 194–195
  postprocessors for, 196–197
  programming in, 197–200
Arcing (see Electrical discharge machines)
Arcs:
  circular interpolation for, 131–132
  preparatory functions for, 101, 104
  programming of, 170–175
Arithmetic APT phase, 195
Arithmetic-logic unit:
  in computers, 40
  in MCUs, 82–83
Artificial intelligence, 30
ASCII coding system, 48–49, 51
  for CNC systems, 85
  for DNC machines, 335–336
  as standard, 94
Assembly drawings, tool, 151
Assembly numbers, tool, 236
Auto power recovery for EDM, 306
Automated tape-controlled tailstocks, 257–259
Automatic operations, 89
  centering, 314–316
  edging, 312–314
  programmable, 332–334
  wire feeding, 305

Automatic Programming of Tools language, 191–192
  general processor for, 194–195
  postprocessor for, 196–197
  programming in, 197–200
Axes, 7–10
  dimensions for, 135
  inversion of, 87
  for machining centers, 222
  motion controls on, 106–111, 303
  presets for, 146–147
  selection function for, 102

Babbage, Charles, 27
Bar feeders, 275
Basic tool angles, 204
BCD (binary-coded decimal) system:
  MCU decoding of, 80–81
  with punched-tape, 45, 48, 50
Beds:
  for chucking and turning centers, 256–257
  for EDM, 295
  for machining centers, 219
Binary-coded decimal systems:
  MCU decoding of, 80–81
  with punched-tape, 45, 48, 50
Binary number system, 4–7
Bits, 35
Blocks, punched-tape, 52
Boring cycle, 99–101
  preparatory functions for, 240–241